T0401968

Baoding Liu

Theory and Practice of Uncertain Programming

Studies in Fuzziness and Soft Computing, Volume 239

Editor-in-Chief

Prof. Janusz Kacprzyk
Systems Research Institute
Polish Academy of Sciences
ul. Newelska 6
01-447 Warsaw
Poland
E-mail: kacprzyk@ibspan.waw.pl

Further volumes of this series can be found on our homepage: springer.com

Vol. 224. Rafael Bello, Rafael Falcón, Witold Pedrycz, Janusz Kacprzyk (Eds.)
Contributions to Fuzzy and Rough Sets Theories and Their Applications, 2008
ISBN 978-3-540-76972-9

Vol. 225. Terry D. Clark, Jennifer M. Larson, John N. Mordeson, Joshua D. Potter, Mark J. Wierman
Applying Fuzzy Mathematics to Formal Models in Comparative Politics, 2008
ISBN 978-3-540-77460-0

Vol. 226. Bhanu Prasad (Ed.)
Soft Computing Applications in Industry, 2008
ISBN 978-3-540-77464-8

Vol. 227. Eugene Roventa, Tiberiu Spircu
Management of Knowledge Imperfection in Building Intelligent Systems, 2008
ISBN 978-3-540-77462-4

Vol. 228. Adam Kasperski
Discrete Optimization with Interval Data, 2008
ISBN 978-3-540-78483-8

Vol. 229. Sadaaki Miyamoto, Hidetomo Ichihashi, Katsuhiro Honda
Algorithms for Fuzzy Clustering, 2008
ISBN 978-3-540-78736-5

Vol. 230. Bhanu Prasad (Ed.)
Soft Computing Applications in Business, 2008
ISBN 978-3-540-79004-4

Vol. 231. Michal Baczynski, Balasubramaniam Jayaram
Soft Fuzzy Implications, 2008
ISBN 978-3-540-69080-1

Vol. 232. Eduardo Massad, Neli Regina Siqueira Ortega, Laécio Carvalho de Barros, Claudio José Struchiner
Fuzzy Logic in Action: Applications in Epidemiology and Beyond, 2008
ISBN 978-3-540-69092-4

Vol. 233. Cengiz Kahraman (Ed.)
Fuzzy Engineering Economics with Applications, 2008
ISBN 978-3-540-70809-4

Vol. 234. Eyal Kolman, Michael Margaliot
Knowledge-Based Neurocomputing: A Fuzzy Logic Approach, 2009
ISBN 978-3-540-88076-9

Vol. 235. Kofi Kissi Dompere
Fuzzy Rationality, 2009
ISBN 978-3-540-88082-0

Vol. 236. Kofi Kissi Dompere
Epistemic Foundations of Fuzziness, 2009
ISBN 978-3-540-88084-4

Vol. 237. Kofi Kissi Dompere
Fuzziness and Approximate Reasoning, 2009
ISBN 978-3-540-88086-8

Vol. 238. Atanu Sengupta, Tapan Kumar Pal
Fuzzy Preference Ordering of Interval Numbers in Decision Problems, 2009
ISBN 978-3-540-89914-3

Vol. 239. Baoding Liu
Theory and Practice of Uncertain Programming, 2009
ISBN 978-3-540-89483-4

Baoding Liu

Theory and Practice of Uncertain Programming

2nd Edition

Author

Baoding Liu
Uncertainty Theory Laboratory
Department of Mathematical Sciences
Tsinghua University
Beijing 100084
China
E-Mail: liu@tsinghua.edu.cn

ISBN 978-3-540-89483-4 e-ISBN 978-3-540-89484-1

DOI 10.1007/978-3-540-89484-1

Studies in Fuzziness and Soft Computing ISSN 1434-9922

Library of Congress Control Number: 2008941016

The first edition appeared as volume 102 of Studies in Fuzziness and Soft Computing ISBN 3-7908-1490-3 Physica-Verlag Heidelberg, 2002.

Typeset & Cover Design: Scientific Publishing Services Pvt. Ltd., Chennai, India.

Printed in acid-free paper

9 8 7 6 5 4 3 2 1

springer.com

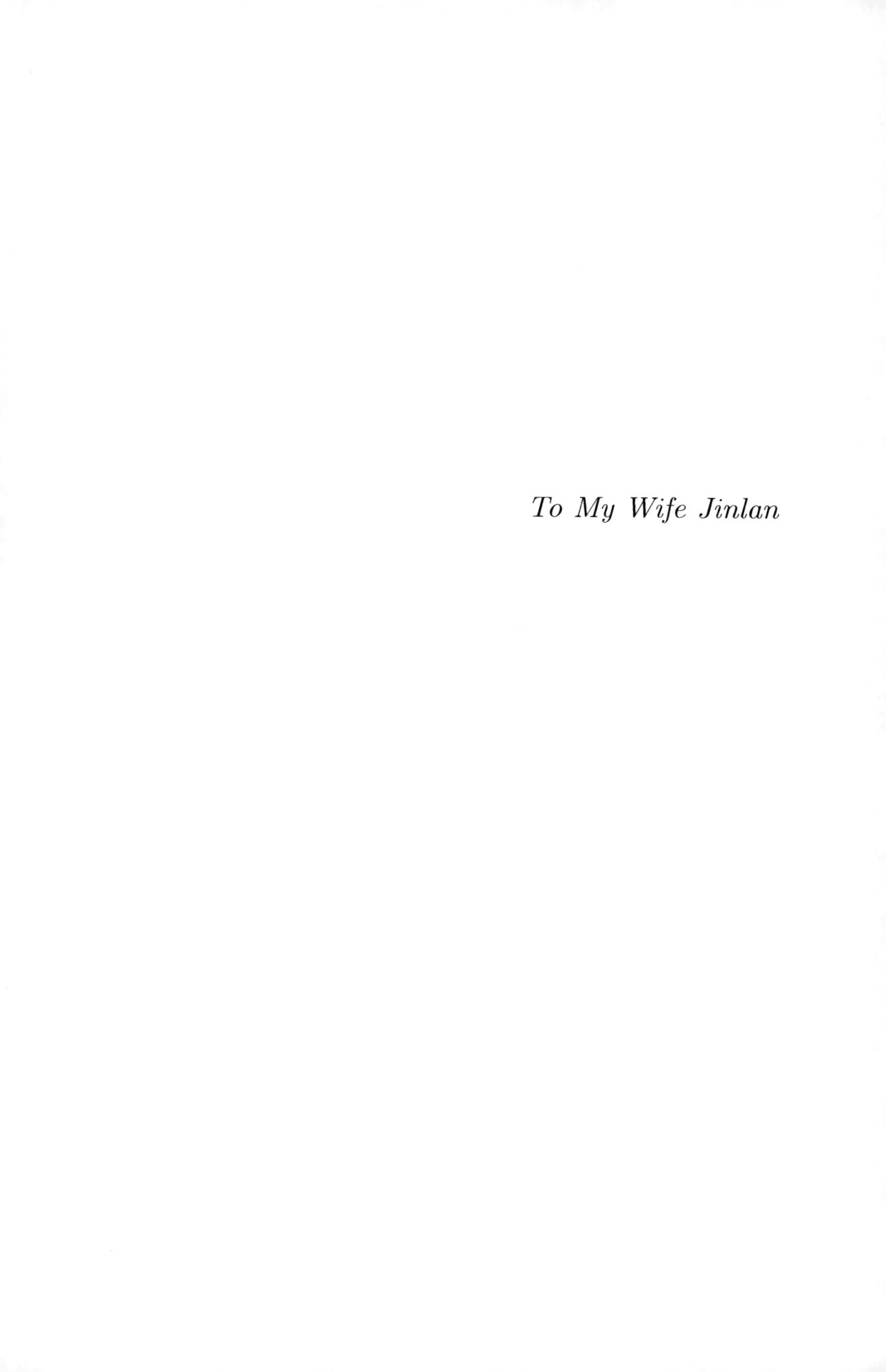

To My Wife Jinlan

Preface

Real-life decisions are usually made in the state of uncertainty. How do we model optimization problems in uncertain environments? How do we solve these models? The main purpose of the book is just to provide uncertain programming theory to answer these questions.

By uncertain programming we mean the optimization theory in uncertain environments. Stochastic programming, fuzzy programming and hybrid programming are subtopics of uncertain programming.

This book provides a self-contained, comprehensive and up-to-date presentation of uncertain programming theory, including numerous modeling ideas and applications in system reliability design, project scheduling problem, vehicle routing problem, facility location problem, and machine scheduling problem.

Numerous intelligent algorithms such as genetic algorithms and neural networks have been developed by researchers of different backgrounds. A natural idea is to integrate these intelligent algorithms to produce more effective and powerful algorithms. In order to solve uncertain programming models, a spectrum of hybrid intelligent algorithms are documented in the book. The author also maintains a website at http://orsc.edu.cn/liu to post the C++ source files of simulations, genetic algorithms, neural networks, and hybrid intelligent algorithms.

For this new edition the entire text has been totally rewritten. More importantly, hybrid variable and hybrid programming are completely new.

It is assumed that readers are familiar with the basic concepts of mathematical programming, and elementary knowledge of C++ language. In order to make the book more readable, some background topics that will be useful in reading the book are also presented. The book is suitable for researchers, engineers, and students in the field of operations research, information science, management science, system science, computer science, and engineering. The readers will learn numerous new modeling ideas and effective algorithms, and find this work a stimulating and useful reference.

Acknowledgment

I am indebted to a series of grants from National Natural Science Foundation, Ministry of Education, and Ministry of Science and Technology of China. I would also like to thank Professor Janusz Kacprzyk for the invitation to publish this book in his series, and Dr. Thomas Ditzinger of Springer for his wonderful cooperation and helpful comments.

September 2008

Baoding Liu
Tsinghua University
http://orsc.edu.cn/liu

Contents

Chapter 1
Mathematical Programming

As one of the most widely used techniques in operations research, *mathematical programming* is defined as a means of maximizing a quantity known as *objective function*, subject to a set of constraints represented by equations and inequalities. Some known subtopics of mathematical programming are linear programming, nonlinear programming, multiobjective programming, goal programming, dynamic programming, and multilevel programming.

It is impossible to cover in a single chapter every concept of mathematical programming. This chapter introduces only the basic concepts and techniques of mathematical programming such that readers gain an understanding of them throughout the book.

1.1 Single-Objective Programming

The general form of single-objective programming (SOP) is written as follows,

$$\begin{cases} \max\ f(\boldsymbol{x}) \\ \text{subject to:} \\ \qquad g_j(\boldsymbol{x}) \le 0, \quad j = 1, 2, \cdots, p \end{cases} \tag{1.1}$$

which maximizes a real-valued function f of $\boldsymbol{x} = (x_1, x_2, \cdots, x_n)$ subject to a set of constraints $g_j(\boldsymbol{x}) \le 0$, $j = 1, 2, \cdots, p$.

Definition 1.1. *In SOP (1.1), we call $\boldsymbol{x}$ a decision vector, and $x_1, x_2, \cdots, x_n$ decision variables. The function f is called the objective function. The set*

$$S = \left\{\boldsymbol{x} \in \Re^n \mid g_j(\boldsymbol{x}) \le 0,\ j = 1, 2, \cdots, p\right\} \tag{1.2}$$

is called the feasible set. An element $\boldsymbol{x}$ in S is called a feasible solution.

Definition 1.2. *A feasible solution $\boldsymbol{x}^*$ is called the optimal solution of SOP (1.1) if and only if*

$$f(\boldsymbol{x}^*) \ge f(\boldsymbol{x}) \tag{1.3}$$

for any feasible solution $\boldsymbol{x}$.

B. Liu: Theory and Practice of Uncertain Programming, STUDFUZZ 239, pp. 1–8.
springerlink.com

One of the outstanding contributions to mathematical programming was known as the Kuhn-Tucker conditions. In order to introduce them, let us give some definitions. An inequality constraint $g_j(\boldsymbol{x}) \leq 0$ is said to be active at a point $\boldsymbol{x}^*$ if $g_j(\boldsymbol{x}^*) = 0$. A point $\boldsymbol{x}^*$ satisfying $g_j(\boldsymbol{x}^*) \leq 0$ is said to be regular if the gradient vectors $\nabla g_j(\boldsymbol{x})$ of all active constraints are linearly independent.

Let $\boldsymbol{x}^*$ be a regular point of the constraints of SOP (1.1) and assume that all the functions $f(\boldsymbol{x})$ and $g_j(\boldsymbol{x}), j = 1, 2, \cdots, p$ are differentiable. If $\boldsymbol{x}^*$ is a local optimal solution, then there exist Lagrange multipliers $\lambda_j, j = 1, 2, \cdots, p$ such that the following Kuhn-Tucker conditions hold,

$$\begin{cases} \nabla f(\boldsymbol{x}^*) - \sum\limits_{j=1}^{p} \lambda_j \nabla g_j(\boldsymbol{x}^*) = 0 \\ \lambda_j g_j(\boldsymbol{x}^*) = 0, \quad j = 1, 2, \cdots, p \\ \lambda_j \geq 0, \quad j = 1, 2, \cdots, p. \end{cases} \tag{1.4}$$

If all the functions $f(\boldsymbol{x})$ and $g_j(\boldsymbol{x}), j = 1, 2, \cdots, p$ are convex and differentiable, and the point $\boldsymbol{x}^*$ satisfies the Kuhn-Tucker conditions (1.4), then it has been proved that the point $\boldsymbol{x}^*$ is a global optimal solution of SOP (1.1).

Linear Programming

If the functions $f(\boldsymbol{x}), g_j(\boldsymbol{x}), j = 1, 2, \cdots, p$ are all linear, then SOP (1.1) is called a *linear programming*.

The feasible set of linear programming is always convex. A point $\boldsymbol{x}$ is called an extreme point of convex set S if $\boldsymbol{x} \in S$ and $\boldsymbol{x}$ cannot be expressed as a convex combination of two points in S. It has been shown that the optimal solution to linear programming corresponds to an extreme point of its feasible set provided that the feasible set S is bounded. This fact is the basis of the *simplex algorithm* which was developed by Dantzig [52] as a very efficient method for solving linear programming.

Roughly speaking, the simplex algorithm examines only the extreme points of the feasible set, rather than all feasible points. At first, the simplex algorithm selects an extreme point as the initial point. The successive extreme point is selected so as to improve the objective function value. The procedure is repeated until no improvement in objective function value can be made. The last extreme point is the optimal solution.

Nonlinear Programming

If at least one of the functions $f(\boldsymbol{x}), g_j(\boldsymbol{x}), j = 1, 2, \cdots, p$ is nonlinear, then SOP (1.1) is called a *nonlinear programming*.

A large number of classical optimization methods have been developed to treat special-structural nonlinear programming based on the mathematical theory concerned with analyzing the structure of problems.

Now we consider a nonlinear programming which is confronted solely with maximizing a real-valued function with domain $\Re^n$. Whether derivatives are available or not, the usual strategy is first to select a point in $\Re^n$ which is thought to be the most likely place where the maximum exists. If there is no information available on which to base such a selection, a point is chosen at random. From this first point an attempt is made to construct a sequence of points, each of which yields an improved objective function value over its predecessor. The next point to be added to the sequence is chosen by analyzing the behavior of the function at the previous points. This construction continues until some termination criterion is met. Methods based upon this strategy are called *ascent methods*, which can be classified as *direct methods*, *gradient methods*, and *Hessian methods* according to the information about the behavior of objective function f. Direct methods require only that the function can be evaluated at each point. Gradient methods require the evaluation of first derivatives of f. Hessian methods require the evaluation of second derivatives. In fact, there is no superior method for all problems. The efficiency of a method is very much dependent upon the objective function.

Integer Programming

Integer programming is a special mathematical programming in which all of the variables are assumed to be only integer values. When there are not only integer variables but also conventional continuous variables, we call it *mixed integer programming.* If all the variables are assumed either 0 or 1, then the problem is termed a *zero-one programming.* Although integer programming can be solved by an *exhaustive enumeration* theoretically, it is impractical to solve realistically sized integer programming problems. The most successful algorithm so far found to solve integer programming is called the *branch-and-bound enumeration* developed by Balas (1965) and Dakin (1965). The other technique to integer programming is the *cutting plane method* developed by Gomory (1959).

1.2 Multiobjective Programming

SOP is related to maximizing a single function subject to a number of constraints. However, it has been increasingly recognized that many real-world decision-making problems involve multiple, noncommensurable, and conflicting objectives which should be considered simultaneously. As an extension, *multiobjective programming* (MOP) is defined as a means of optimizing multiple objective functions subject to a number of constraints, i.e.,

$$\begin{cases} \max\ [f_1(\boldsymbol{x}), f_2(\boldsymbol{x}), \cdots, f_m(\boldsymbol{x})] \\ \text{subject to:} \\ \qquad g_j(\boldsymbol{x}) \le 0,\ j = 1, 2, \cdots, p \end{cases} \tag{1.5}$$

where $f_i(\boldsymbol{x})$ are objective functions, $i = 1, 2, \cdots, m$, and $g_j(\boldsymbol{x}) \le 0$ are system constraints, $j = 1, 2, \cdots, p$.

When the objectives are in conflict, there is no optimal solution that simultaneously maximizes all the objective functions. For this case, we employ a concept of *Pareto solution*, which means that it is impossible to improve any one objective without sacrificing on one or more of the other objectives.

Definition 1.3. *A feasible solution $\boldsymbol{x}^*$ is said to be a Pareto solution if there is no feasible solution $\boldsymbol{x}$ such that*

$$f_i(\boldsymbol{x}) \ge f_i(\boldsymbol{x}^*), \quad i = 1, 2, \cdots, m \tag{1.6}$$

and $f_j(\boldsymbol{x}) > f_j(\boldsymbol{x}^*)$ for at least one index j.

If the decision maker has a real-valued *preference function* aggregating the m objective functions, then we may maximize the aggregating preference function subject to the same set of constraints. This model is referred to as a *compromise model* whose solution is called a *compromise solution.*

The first well-known compromise model is set up by weighting the objective functions, i.e.,

$$\begin{cases} \max \sum_{i=1}^{m} \lambda_i f_i(\boldsymbol{x}) \\ \text{subject to:} \\ \qquad g_j(\boldsymbol{x}) \le 0, \ j = 1, 2, \cdots, p \end{cases} \tag{1.7}$$

where the weights $\lambda_1, \lambda_2, \cdots, \lambda_m$ are nonnegative numbers with $\lambda_1 + \lambda_2 + \cdots + \lambda_m = 1$. Note that the solution of (1.7) must be a Pareto solution of the original one.

The second way is related to minimizing the *distance function* from a solution $(f_1(\boldsymbol{x}), f_2(\boldsymbol{x}), \cdots, f_m(\boldsymbol{x}))$ to an ideal vector $(f_1^*, f_2^*, \cdots, f_m^*)$, where f_i^* are the optimal values of the ith objective functions without considering other objectives, $i = 1, 2, \cdots, m$, respectively, i.e.,

$$\begin{cases} \min \sqrt{(f_1(\boldsymbol{x}) - f_1^*)^2 + \cdots + (f_m(\boldsymbol{x}) - f_m^*)^2} \\ \text{subject to:} \\ \qquad g_j(\boldsymbol{x}) \le 0, \ j = 1, 2, \cdots, p. \end{cases} \tag{1.8}$$

By the third way a compromise solution can be found via an *interactive approach* consisting of a sequence of decision phases and computation phases. Various interactive approaches have been developed in the past literature.

1.3 Goal Programming

Goal programming (GP) was developed by Charnes and Cooper [38] and subsequently studied by many researchers. GP can be regarded as a special compromise model for multiobjective programming and has been applied in a wide variety of real-world problems.

In multiobjective decision-making problems, we assume that the decision-maker is able to assign a target level for each goal and the key idea is to minimize the deviations (positive, negative, or both) from the target levels. In the real-world situation, the goals are achievable only at the expense of other goals and these goals are usually incompatible. Therefore, there is a need to establish a hierarchy of importance among these incompatible goals so as to satisfy as many goals as possible in the order specified. The general form of GP is written as follows,

$$\begin{cases} \min \sum_{j=1}^{l} P_j \sum_{i=1}^{m} (u_{ij} d_i^+ \vee 0 + v_{ij} d_i^- \vee 0) \\ \text{subject to:} \\ \qquad f_i(\boldsymbol{x}) - b_i = d_i^+, \ i = 1, 2, \cdots, m \\ \qquad b_i - f_i(\boldsymbol{x}) = d_i^-, \ i = 1, 2, \cdots, m \\ \qquad g_j(\boldsymbol{x}) \le 0, \qquad j = 1, 2, \cdots, p \end{cases} \tag{1.9}$$

where P_j is the preemptive priority factor which expresses the relative importance of various goals, $P_j \gg P_{j+1}$, for all j, u_{ij} is the weighting factor corresponding to positive deviation for goal i with priority j assigned, v_{ij} is the weighting factor corresponding to negative deviation for goal i with priority j assigned, $d_i^+ \vee 0$ (i.e., the maximum value of d_i^+ and 0) is the positive deviation from the target of goal i, $d_i^- \vee 0$ (i.e., the maximum value of d_i^- and 0) is the negative deviation from the target of goal i, f_i is a function in goal constraints, g_j is a function in system constraints, b_i is the target value according to goal i, l is the number of priorities, m is the number of goal constraints, and p is the number of system constraints. Sometimes, the objective function of GP (1.9) is written as

$$\text{lexmin} \left\{ \sum_{i=1}^{m} (u_{i1} d_i^+ \vee 0 + v_{i1} d_i^- \vee 0), \cdots, \sum_{i=1}^{m} (u_{il} d_i^+ \vee 0 + v_{il} d_i^- \vee 0) \right\}$$

where lexmin represents lexicographically minimizing the objective vector.

Linear GP can be successfully solved by the simplex goal method. The approaches of nonlinear GP are summarized by Saber and Ravindran [270] and the efficiency of these approaches varies. They are classified as follows: (a) simplex-based approach, whose main idea lies in converting the nonlinear GP into a set of approximation linear GPs which can be handled by the simplex goal method; (b) direct search approach [49], in which the given nonlinear GP is translated into a set of SOPs, and then the SOPs are solved by the direct search methods; (c) gradient-based approach [152][270], which utilizes the gradient of constraints to identify a feasible direction and then solves the GP based on the feasible direction method; (d) interactive approach [309][226], which can yield a satisfactory solution in a relatively few iterations since the decision-maker is involved in the solution process; and (e) genetic algorithm [86], which can deal with complex nonlinear GP but have to spend more CPU time.

1.4 Dynamic Programming

Let us denote a multistage decision process by $[\boldsymbol{a}, T(\boldsymbol{a}, \boldsymbol{x})]$, where $\boldsymbol{a}$ is called *state*, $T(\boldsymbol{a}, \boldsymbol{x})$ is called a *state transition function*, and $\boldsymbol{x}$ is called *decision vector*. It is clear that the state transition function depends on both state $\boldsymbol{a}$ and decision vector $\boldsymbol{x}$. We suppose that we have sufficient influence over the process so that at each stage we can choose a decision vector $\boldsymbol{x}$ from the allowable set. Assume that there are N stages in the time horizon, and let $\boldsymbol{x}_i$ be the decision vector at the ith stage. Then we have the following sequence,

$$\boldsymbol{a}_1 = \boldsymbol{a}_0, \quad \text{(an initial state)}$$

$$\boldsymbol{a}_{i+1} = T(\boldsymbol{a}_i, \boldsymbol{x}_i), \quad i = 1, 2, \cdots, N-1.$$

We are concerned with processes in which the decision vectors $\boldsymbol{x}_i$'s are chosen so as to maximize a *criterion function* $R(\boldsymbol{a}_1, \boldsymbol{a}_2, \cdots, \boldsymbol{a}_N; \boldsymbol{x}_1, \boldsymbol{x}_2, \cdots, \boldsymbol{x}_N)$. A decision is called *optimal* if it maximizes the criterion function.

In view of the general nature of the criterion function R, the decision vectors $\boldsymbol{x}_i$'s are dependent upon the current state of the system as well as the past and future states and decisions. However, there are some criterion functions which have some special structures so that the decision is dependent only on the current state. In this special but extremely important case, the optimal policy is characterized by Bellman's principle of optimality: *An optimal policy has the property that whatever the initial state and initial decision are, the remaining decision must constitute an optimal policy with regard to the state resulting from the first decision.*

Fortunately, many important criteria have the vital property of divorcing the past from the present. In general, it is easy to predict this property from the nature of the original multistage decision process. For example, let us consider a problem of maximizing the following special-structured function

$$R(\boldsymbol{a}_1, \boldsymbol{a}_2, \cdots, \boldsymbol{a}_N; \boldsymbol{x}_1, \boldsymbol{x}_2, \cdots, \boldsymbol{x}_N) = \sum_{i=1}^{N} r_i(\boldsymbol{a}_i, \boldsymbol{x}_i) \tag{1.10}$$

subject to $g_i(\boldsymbol{a}_i, \boldsymbol{x}_i) \le 0$ for $i = 1, 2, \cdots, N$. Let $f_n(\boldsymbol{a})$ be the maximum values of criterion function R, starting in state $\boldsymbol{a}$ at the stage n, $n = 1, 2, \cdots, N$, respectively. Then by Bellman's principle of optimality, we have

$$\begin{cases} f_N(\boldsymbol{a}) = \max\limits_{g_N(\boldsymbol{a},\boldsymbol{x})\le 0} r_N(\boldsymbol{a},\boldsymbol{x}) \\ f_{N-1}(\boldsymbol{a}) = \max\limits_{g_{N-1}(\boldsymbol{a},\boldsymbol{x})\le 0} \{r_{N-1}(\boldsymbol{a},\boldsymbol{x}) + f_N(T(\boldsymbol{a},\boldsymbol{x}))\} \\ \cdots \\ f_1(\boldsymbol{a}) = \max\limits_{g_1(\boldsymbol{a},\boldsymbol{x})\le 0} \{r_1(\boldsymbol{a},\boldsymbol{x}) + f_2(T(\boldsymbol{a},\boldsymbol{x}))\}. \end{cases} \tag{1.11}$$

Please mention that,

$$\max_{\boldsymbol{x}_1,\boldsymbol{x}_2,\cdots,\boldsymbol{x}_N} R(\boldsymbol{a}_1,\boldsymbol{a}_2,\cdots,\boldsymbol{a}_N;\boldsymbol{x}_1,\boldsymbol{x}_2,\cdots,\boldsymbol{x}_N) = f_1(\boldsymbol{a}_0). \tag{1.12}$$

The system of equations (1.11) is called *dynamic programming* (DP) by Richard Bellman [12] which can be simply written as

$$\begin{cases} f_N(\boldsymbol{a}) = \max\limits_{g_N(\boldsymbol{a},\boldsymbol{x})\le 0} r_N(\boldsymbol{a},\boldsymbol{x}) \\ f_n(\boldsymbol{a}) = \max\limits_{g_n(\boldsymbol{a},\boldsymbol{x})\le 0} \{r_n(\boldsymbol{a},\boldsymbol{x}) + f_{n+1}(T(\boldsymbol{a},\boldsymbol{x}))\} \\ n \le N-1. \end{cases} \tag{1.13}$$

In order to obtain the optimal solutions in reasonable time for real practical problems, we should develop effectively computational algorithms for DP. To explore the general DP algorithms, readers may consult the book by Bertsekas and Tsitsiklis [16] in which numerous different ways to solve DP problems have been suggested.

1.5 Multilevel Programming

Multilevel programming (MLP) offers a means of studying decentralized decision systems in which we assume that the leader and followers may have their own decision variables and objective functions, and the leader can only influence the reactions of followers through his own decision variables, while the followers have full authority to decide how to optimize their own objective functions in view of the decisions of the leader and other followers.

We now assume that in a decentralized two-level decision system there is one leader and m followers. Let $\boldsymbol{x}$ and $\boldsymbol{y}_i$ be the decision vectors of the leader and the ith followers, $i = 1, 2, \cdots, m$, respectively. We also assume that the objective functions of the leader and ith followers are $F(\boldsymbol{x}, \boldsymbol{y}_1, \cdots, \boldsymbol{y}_m)$ and $f_i(\boldsymbol{x}, \boldsymbol{y}_1, \cdots, \boldsymbol{y}_m)$, $i = 1, 2, \cdots, m$, respectively.

In addition, let the feasible set of control vector $\boldsymbol{x}$ of the leader be determined by

$$G(\boldsymbol{x}) \le 0 \tag{1.14}$$

where G is a vector-valued function of decision vector $\boldsymbol{x}$ and 0 is a vector with zero components. Then for each decision $\boldsymbol{x}$ chosen by the leader, the feasibility of decision vectors $\boldsymbol{y}_i$ of the ith followers should be dependent on not only $\boldsymbol{x}$ but also $\boldsymbol{y}_1, \cdots, \boldsymbol{y}_{i-1}, \boldsymbol{y}_{i+1}, \cdots, \boldsymbol{y}_m$, and generally represented by

$$g_i(\boldsymbol{x}, \boldsymbol{y}_1, \boldsymbol{y}_2, \cdots, \boldsymbol{y}_m) \le 0 \tag{1.15}$$

where g_i are vector-valued functions, $i = 1, 2, \cdots, m$, respectively.

Assume that the leader first chooses his decision vector $\boldsymbol{x}$, and the followers determine their decision array $(\boldsymbol{y}_1, \boldsymbol{y}_2, \cdots, \boldsymbol{y}_m)$ after that. In order to find the optimal decision vector of the leader, we have to use the following bilevel programming,

$$\begin{cases} \max\limits_{\boldsymbol{x}} F(\boldsymbol{x}, \boldsymbol{y}_1^*, \boldsymbol{y}_2^*, \cdots, \boldsymbol{y}_m^*) \\ \text{subject to:} \\ \qquad G(\boldsymbol{x}) \le 0 \\ \qquad (\boldsymbol{y}_1^*, \boldsymbol{y}_2^*, \cdots, \boldsymbol{y}_m^*) \text{ solves problems } (i = 1, 2, \cdots, m) \\ \qquad \begin{cases} \max\limits_{\boldsymbol{y}_i} f_i(\boldsymbol{x}, \boldsymbol{y}_1, \boldsymbol{y}_2, \cdots, \boldsymbol{y}_m) \\ \text{subject to:} \\ \qquad g_i(\boldsymbol{x}, \boldsymbol{y}_1, \boldsymbol{y}_2, \cdots, \boldsymbol{y}_m) \le 0. \end{cases} \end{cases} \tag{1.16}$$

Definition 1.4 *Let $\boldsymbol{x}$ be a fixed decision vector of the leader. A Nash equilibrium of followers with respect to $\boldsymbol{x}$ is the feasible array $(\boldsymbol{y}_1^*, \boldsymbol{y}_2^*, \cdots, \boldsymbol{y}_m^*)$ such that*

$$f_i(\boldsymbol{x}, \boldsymbol{y}_1^*, \cdots, \boldsymbol{y}_{i-1}^*, \boldsymbol{y}_i, \boldsymbol{y}_{i+1}^*, \cdots, \boldsymbol{y}_m^*) \le f_i(\boldsymbol{x}, \boldsymbol{y}_1^*, \cdots, \boldsymbol{y}_{i-1}^*, \boldsymbol{y}_i^*, \boldsymbol{y}_{i+1}^*, \cdots, \boldsymbol{y}_m^*)$$

for any feasible array $(\boldsymbol{y}_1^*, \cdots, \boldsymbol{y}_{i-1}^*, \boldsymbol{y}_i, \boldsymbol{y}_{i+1}^*, \cdots, \boldsymbol{y}_m^*)$ and $i = 1, 2, \cdots, m$.

Definition 1.5 *Suppose that $\boldsymbol{x}^*$ is a feasible decision vector of the leader and $(\boldsymbol{y}_1^*, \boldsymbol{y}_2^*, \cdots, \boldsymbol{y}_m^*)$ is a Nash equilibrium of followers with respect to $\boldsymbol{x}^*$. We call $(\boldsymbol{x}^*, \boldsymbol{y}_1^*, \boldsymbol{y}_2^*, \cdots, \boldsymbol{y}_m^*)$ a Stackelberg-Nash equilibrium to MLP (1.16) if and only if*

$$F(\overline{\boldsymbol{x}}, \overline{\boldsymbol{y}}_1, \overline{\boldsymbol{y}}_2, \cdots, \overline{\boldsymbol{y}}_m) \le F(\boldsymbol{x}^*, \boldsymbol{y}_1^*, \boldsymbol{y}_2^*, \cdots, \boldsymbol{y}_m^*) \tag{1.17}$$

for any feasible $\overline{\boldsymbol{x}}$ and Nash equilibrium $(\overline{\boldsymbol{y}}_1, \overline{\boldsymbol{y}}_2, \cdots, \overline{\boldsymbol{y}}_m)$ with respect to $\overline{\boldsymbol{x}}$.

Ben-Ayed and Blair [14] showed that MLP is an NP-hard problem. In order to solve MLP, a lot of numerical algorithms have been developed, for example, implicit enumeration scheme (Candler and Townsley [34]), the kth best algorithm (Bialas and Karwan [18]), parametric complementary pivot algorithm (Bialas and Karwan [18]), one-dimensional grid search algorithm (Bard [8][10]), branch-and-bound algorithm (Bard and Moore [9]), the steepest-descent direction (Savard and Gauvin [277]), and genetic algorithm (Liu [170]).

Chapter 2
Genetic Algorithms

Genetic algorithm (GA) is a stochastic search method for optimization problems based on the mechanics of natural selection and natural genetics (i.e., survival of the fittest). GA has demonstrated considerable success in providing good solutions to many complex optimization problems and received more and more attentions during the past three decades. When the objective functions to be optimized in the optimization problems are multimodal or the search spaces are particularly irregular, algorithms need to be highly robust in order to avoid getting stuck at a local optimal solution. The advantage of GA is just able to obtain the global optimal solution fairly. In addition, GA does not require the specific mathematical analysis of optimization problems, which makes GA easily coded by users who are not necessarily good at mathematics and algorithms.

One of the important technical terms in GA is *chromosome*, which is usually a string of symbols or numbers. A chromosome is a coding of a solution of an optimization problem, not necessarily the solution itself. GA starts with an initial set of randomly generated chromosomes called a *population*. The number of individuals in the population is a predetermined integer and is called *population size*. All chromosomes are evaluated by the so-called *evaluation function*, which is some measure of *fitness*. A new population will be formed by a *selection process* using some *sampling mechanism* based on the fitness values. The cycle from one population to the next one is called a *generation*. In each new generation, all chromosomes will be updated by the *crossover* and *mutation* operations. The revised chromosomes are also called *offspring*. The selection process selects chromosomes to form a new population and the genetic system enters a new generation. After performing the genetic system a given number of cycles, we decode the best chromosome into a solution which is regarded as the optimal solution of the optimization problem.

GA has been well-documented in the literature, such as in Holland [96], Goldberg [89], Michalewicz [230], Fogel [69], Koza [137][138], Liu [181], and have been applied to a wide variety of problems. The aim of this section is to introduce an effective GA for solving complex optimization problems.

B. Liu: Theory and Practice of Uncertain Programming, STUDFUZZ 239, pp. 9–17.
springerlink.com

Moreover, we design this algorithm for solving not only single-objective optimization but also multiobjective programming, goal programming, and multilevel programming. Finally, we illustrate the effectiveness of GA by some numerical examples.

2.1 Representation Structure

A key problem of GA is how to encode a solution $\boldsymbol{x} = (x_1, x_2, \cdots, x_n)$ into a chromosome $V = (v_1, v_2, \cdots, v_m)$. That is, we must construct a link between a solution space and a coding space. The mapping from the solution space to coding space is called *encoding*. The mapping from the coding space to solution space is called *decoding*.

It is clear that the representation structure is problem-dependent. For example, let (x_1, x_2, x_3) be a solution vector in the solution space

$$\begin{cases} x_1 + x_2^2 + x_3^3 = 1 \\ x_1 \geq 0,\ x_2 \geq 0,\ x_3 \geq 0. \end{cases} \tag{2.1}$$

We may encode the solution by a chromosome (v_1, v_2, v_3) in the coding space

$$v_1 \geq 0, \quad v_2 \geq 0, \quad v_3 \geq 0. \tag{2.2}$$

Then the encoding and decoding processes are determined by the link

$$x_1 = \frac{v_1}{v_1 + v_2 + v_3}, \quad x_2 = \sqrt{\frac{v_2}{v_1 + v_2 + v_3}}, \quad x_3 = \sqrt[3]{\frac{v_3}{v_1 + v_2 + v_3}}. \tag{2.3}$$

2.2 Handling Constraints

In mathematical programming, if there are some equality constraints, for example,

$$h_k(\boldsymbol{x}) = 0, \quad k = 1, 2, \cdots, q, \tag{2.4}$$

we should eliminate the q equality constraints by replacing q variables of them with the representation of the remaining variables, where the representation is obtained by solving the system of equalities in the constraints.

If we cannot do so, we may eliminate the equality constraints by Lagrangian method based on the idea of transforming a constrained problem into an unconstrained one.

2.3 Initialization Process

We define an integer *pop_size* as the number of chromosomes and initialize *pop_size* chromosomes randomly. Usually, it is difficult for complex optimization problems to produce feasible chromosomes explicitly.

Assume that the decision-maker can predetermine a region which contains the optimal solution (not necessarily the whole feasible set). Such a region is also problem-dependent. At any rate, the decision-maker can provide such a region, only it may be a bit too large. Usually, this region will be designed to have a nice sharp, for example, a hypercube, because the computer can easily sample points from a hypercube.

We generate a random point from the hypercube and check the feasibility of this point. If it is feasible, then it will be accepted as a chromosome. If not, then we regenerate a point from the hypercube randomly until a feasible one is obtained. We can make *pop_size* initial feasible chromosomes

$$V_1, V_2, \cdots, V_{pop\_size}$$

by repeating the above process *pop_size* times.

2.4 Evaluation Function

Evaluation function, denoted by $Eval(V)$, is to assign a probability of reproduction to each chromosome V so that its likelihood of being selected is proportional to its fitness relative to the other chromosomes in the population. That is, the chromosomes with higher fitness will have more chance to produce offspring by using *roulette wheel selection*.

Let $V_1, V_2, \cdots, V_{pop\_size}$ be the *pop_size* chromosomes at the current generation. One well-known evaluation function is based on allocation of reproductive trials according to rank rather than actual objective values. No matter what type of mathematical programming it is, it is reasonable to assume that the decision-maker can give an order relationship among the *pop_size* chromosomes $V_1, V_2, \cdots, V_{pop\_size}$ such that the *pop_size* chromosomes can be rearranged from good to bad (i.e., the better the chromosome is, the smaller the ordinal number it has). For example, for a single-objective maximizing problem, a chromosome with larger objective value is better; for a multiobjective programming, we may define a preference function to evaluate the chromosomes; for a goal programming, we have the following order relationship for the chromosomes: for any two chromosomes, if the higher-priority objectives are equal to each other, then, in the current priority level, the one with minimal objective value is better. If two different chromosomes have the same objective values at every level, then we are indifferent between them. For this case, we rearrange them randomly.

Now let a parameter $a \in (0, 1)$ in the genetic system be given (for example, $a = 0.05$). We can define the *rank-based evaluation function* as follows,

$$Eval(V_i) = a(1-a)^{i-1}, \qquad i = 1, 2, \cdots, pop\_size. \tag{2.5}$$

Note that $i = 1$ means the best individual, $i = pop\_size$ the worst one.

2.5 Selection Process

The selection process is based on spinning the roulette wheel *pop_size* times. Each time we select a single chromosome for a new population. The roulette wheel is a fitness-proportional selection. No matter what type of evaluation function is employed, the selection process is always stated as follows:

Algorithm 2.1 (Selection Process)
Step 1. Calculate the cumulative probability q_i for each chromosome V_i,

$$q_0 = 0, \quad q_i = \sum_{j=1}^{i} Eval(V_j), \quad i = 1, 2, \cdots, pop\_size.$$

Step 2. Generate a random number r in $(0, q_{pop\_size}]$.
Step 3. Select the chromosome V_i such that $q_{i-1} < r \le q_i$.
Step 4. Repeat the second and third steps *pop_size* times and obtain *pop_size* copies of chromosome.

Please note that in the above selection process we do not require the condition $q_{pop\_size} = 1$. In fact, if we want, we can divide all q_i's, $i = 1, 2, \cdots, pop\_size$, by q_{pop_size} such that $q_{pop\_size} = 1$ and the new probabilities are also proportional to the fitnesses. However, it does not exert any influence on the genetic process.

2.6 Crossover Operation

We define a parameter P_c of a genetic system as the probability of crossover. This probability gives us the expected number $P_c \cdot pop\_size$ of chromosomes undergoing the crossover operation.

In order to determine the parents for crossover operation, let us do the following process repeatedly from $i = 1$ to *pop_size*: generating a random number r from the interval $[0, 1]$, the chromosome V_i is selected as a parent if $r < P_c$. We denote the selected parents by $V_1', V_2', V_3', \cdots$ and divide them into the following pairs:

$$(V_1', V_2'), \quad (V_3', V_4'), \quad (V_5', V_6'), \quad \cdots$$

Let us illustrate the crossover operator on each pair by (V_1', V_2'). At first, we generate a random number c from the open interval $(0, 1)$, then the crossover operator on V_1' and V_2' will produce two children X and Y as follows:

$$X = c \cdot V_1' + (1 - c) \cdot V_2', \qquad Y = (1 - c) \cdot V_1' + c \cdot V_2'. \tag{2.6}$$

If the feasible set is convex, this crossover operation ensures that both children are feasible if both parents are. However, in many cases, the feasible set is not necessarily convex, nor is it hard to verify the convexity. Thus we must check the feasibility of each child before accepting it. If both children are feasible, then we replace the parents with them. If not, we keep the feasible one if it exists, and then redo the crossover operator by regenerating a random number c until two feasible children are obtained or a given number of cycles is finished. In this case, we only replace the parents with the feasible children.

2.7 Mutation Operation

We define a parameter P_m of a genetic system as the probability of mutation. This probability gives us the expected number of $P_m \cdot pop\_size$ of chromosomes undergoing the mutation operations.

In a similar manner to the process of selecting parents for crossover operation, we repeat the following steps from $i = 1$ to *pop_size*: generating a random number r from the interval $[0, 1]$, the chromosome V_i is selected as a parent for mutation if $r < P_m$.

For each selected parent, denoted by $V = (v_1, v_2, \cdots, v_m)$, we mutate it in the following way. Let M be an appropriate large positive number. We choose a mutation direction $\boldsymbol{d}$ in $\Re^m$ randomly. If $V + M \cdot \boldsymbol{d}$ is not feasible, then we set M as a random number between 0 and M until it is feasible. If the above process cannot find a feasible solution in a predetermined number of iterations, then we set $M = 0$. Anyway, we replace the parent V with its child

$$X = V + M \cdot \boldsymbol{d}. \tag{2.7}$$

2.8 General Procedure

Following selection, crossover, and mutation, the new population is ready for its next evaluation. GA will terminate after a given number of cyclic repetitions of the above steps or a suitable solution has been found. We now summarize the GA for optimization problems as follows.

Algorithm 2.2 (Genetic Algorithm)
Step 1. Initialize *pop_size* chromosomes at random.
Step 2. Update the chromosomes by crossover and mutation operations.
Step 3. Calculate the objective values for all chromosomes.
Step 4. Compute the fitness of each chromosome via the objective values.
Step 5. Select the chromosomes by spinning the roulette wheel.
Step 6. Repeat the second to fifth steps for a given number of cycles.
Step 7. Report the best chromosome as the optimal solution.

Remark 2.1. It is well-known that the best chromosome does not necessarily appear in the last generation. Thus we have to keep the best one from the beginning. If we find a better one in the new population, then we replace the old one with it. This chromosome will be reported as the optimal solution after finishing the evolution.

2.9 Numerical Experiments

Example 2.1. Now we use GA to solve the following maximization problem,

$$\begin{cases} \max \sqrt{x_1} + \sqrt{x_2} + \sqrt{x_3} \\ \text{subject to:} \\ \qquad x_1^2 + 2x_2^2 + 3x_3^2 \le 1 \\ \qquad x_1, x_2, x_3 \ge 0. \end{cases} \tag{2.8}$$

We may encode a solution $\boldsymbol{x} = (x_1, x_2, x_3)$ into a chromosome $V = (v_1, v_2, v_3)$, and decode the chromosome into the solution in the following way,

$$x_1 = v_1, \quad x_2 = v_2, \quad x_3 = v_3.$$

It is easy to know that the feasible coding space is contained in the following hypercube

$$\mathcal{V} = \left\{(v_1, v_2, v_3) \;\middle|\; 0 \le v_1 \le 1,\ 0 \le v_2 \le 1,\ 0 \le v_3 \le 1\right\}$$

which is simple for the computer because we can easily sample points from it. We may take

$$v_1 = \mathcal{U}(0,1), \quad v_2 = \mathcal{U}(0,1), \quad v_3 = \mathcal{U}(0,1) \tag{2.9}$$

where the function $\mathcal{U}(a,b)$ generates uniformly distributed variables on the interval $[a,b]$ and will be discussed in detail later. If this chromosome is infeasible, then we reject it and regenerate one by (2.9). If the generated chromosome is feasible, then we accept it as one in the population. After finite times, we can obtain 30 feasible chromosomes. A run of GA with 400 generations shows that the optimal solution is

$$(x_1^*, x_2^*, x_3^*) = (0.636, 0.395, 0.307)$$

whose objective value is 1.980.

Example 2.2. GA is also able to solve the following nonlinear goal programming,

$$
\begin{cases}
\text{lexmin}\,\{d_1^- \vee 0, d_2^- \vee 0, d_3^- \vee 0\} \\
\text{subject to:} \\
\qquad 3 - \sqrt{x_1} = d_1^- \\
\qquad 4 - \sqrt{x_1 + 2x_2} = d_2^- \\
\qquad 5 - \sqrt{x_1 + 2x_2 + 3x_3} = d_3^- \\
\qquad x_1^2 + x_2^2 + x_3^2 \le 100 \\
\qquad x_1, x_2, x_3 \ge 0.
\end{cases}
$$

We may encode a solution $\boldsymbol{x} = (x_1, x_2, x_3)$ into a chromosome $V = (v_1, v_2, v_3)$, and decode the chromosome into the solution in the following way,

$$x_1 = v_1, \quad x_2 = v_2, \quad x_3 = v_3.$$

Since the feasible coding space is contained in the following hypercube

$$\mathcal{V} = \left\{(v_1, v_2, v_3) \,\middle|\, 0 \le v_1 \le 10,\ 0 \le v_2 \le 10,\ 0 \le v_3 \le 10\right\},$$

we may take $v_1 = \mathcal{U}(0,10)$, $v_2 = \mathcal{U}(0,10)$, and $v_3 = \mathcal{U}(0,10)$, and accept it as a chromosome if it is feasible. It is clear that we can make 30 feasible chromosomes in finite times. A run of GA with 2000 generations shows that the optimal solution is

$$(x_1^*, x_2^*, x_3^*) = (9.000, 3.500, 2.597)$$

which satisfies the first two goals, but the last objective is 0.122.

Example 2.3. For the following bilevel programming model,

$$
\begin{cases}
\max\limits_{\boldsymbol{x}} F(\boldsymbol{x}, \boldsymbol{y}_1^*, \boldsymbol{y}_2^*, \cdots, \boldsymbol{y}_m^*) \\
\text{subject to:} \\
\qquad G(\boldsymbol{x}) \le 0 \\
\qquad (\boldsymbol{y}_1^*, \boldsymbol{y}_2^*, \cdots, \boldsymbol{y}_m^*) \text{ solves problems } (i = 1, 2, \cdots, m) \\
\qquad \begin{cases}
\max\limits_{\boldsymbol{y}_i} f_i(\boldsymbol{x}, \boldsymbol{y}_1, \boldsymbol{y}_2, \cdots, \boldsymbol{y}_m) \\
\text{subject to:} \\
\qquad g_i(\boldsymbol{x}, \boldsymbol{y}_1, \boldsymbol{y}_2, \cdots, \boldsymbol{y}_m) \le 0,
\end{cases}
\end{cases}
\tag{2.10}
$$

we define symbols

$$\boldsymbol{y}_{-i} = (\boldsymbol{y}_1, \cdots, \boldsymbol{y}_{i-1}, \boldsymbol{y}_{i+1}, \cdots, \boldsymbol{y}_m), \quad i = 1, 2, \cdots, m. \tag{2.11}$$

For any decision $\boldsymbol{x}$ revealed by the leader, if the ith follower knows the strategies $\boldsymbol{y}_{-i}$ of other followers, then the optimal reaction of the ith follower is represented by a mapping $\boldsymbol{y}_i = r_i(\boldsymbol{y}_{-i})$, which should solve the subproblem

$$\begin{cases} \max\limits_{\boldsymbol{y}_i} f_i(\boldsymbol{x}, \boldsymbol{y}_1, \boldsymbol{y}_2, \cdots, \boldsymbol{y}_m) \\ \text{subject to:} \\ \qquad g_i(\boldsymbol{x}, \boldsymbol{y}_1, \boldsymbol{y}_2, \cdots, \boldsymbol{y}_m) \le 0. \end{cases} \tag{2.12}$$

In order to search for the Stackelberg-Nash equilibrium, Liu [170] designed a GA to solve multilevel programming. We first compute the Nash equilibrium with respect to any decision revealed by the leader. It is clear that the Nash equilibrium of the m followers will be the solution $(\boldsymbol{y}_1, \boldsymbol{y}_2, \cdots, \boldsymbol{y}_m)$ of the system of equations

$$\boldsymbol{y}_i = r_i(\boldsymbol{y}_{-i}), \quad i = 1, 2, \cdots, m. \tag{2.13}$$

In other words, we should find a fixed point of the vector-valued function $(r_1, r_2, \cdots, r_m)$. In order to solve the system of equations (2.13), we should design some efficient algorithms. The argument breaks down into three cases.

(a) If we have explicit expressions of all functions r_i, $i = 1, 2, \cdots, m$, then we might get an analytic solution to the system (2.13). Unfortunately, it is almost impossible to do this in practice.

(b) The system (2.13) might be solved by some iterative method that generates a sequence of points $\boldsymbol{y}^k = (\boldsymbol{y}_1^k, \boldsymbol{y}_2^k, \cdots, \boldsymbol{y}_m^k)$, $k = 0, 1, 2, \cdots$ via the iteration formula

$$\boldsymbol{y}_i^{k+1} = r_i(\boldsymbol{y}_{-i}^k), \quad i = 1, 2, \cdots, m \tag{2.14}$$

where $\boldsymbol{y}_{-i}^k = (\boldsymbol{y}_1^k, \cdots, \boldsymbol{y}_{i-1}^k, \boldsymbol{y}_{i+1}^k, \cdots, \boldsymbol{y}_m^k)$. However, generally speaking, it is not easy to verify the conditions on the convergence of the iterative method for practical problems.

(c) If the iterative method fails to find a fixed point, we may employ GA to solve the following minimization problem,

$$\min R(\boldsymbol{y}_1, \boldsymbol{y}_2, \cdots, \boldsymbol{y}_m) = \sum_{i=1}^{m} \|\boldsymbol{y}_i - r_i(\boldsymbol{y}_{-i})\| \tag{2.15}$$

If an array $(\boldsymbol{y}_1^*, \boldsymbol{y}_2^*, \cdots, \boldsymbol{y}_m^*)$ makes $R(\boldsymbol{y}_1^*, \boldsymbol{y}_2^*, \cdots, \boldsymbol{y}_m^*) = 0$, then $\boldsymbol{y}_i^* = r_i(\boldsymbol{y}_{-i}^*)$, $i = 1, 2, \cdots, m$ and $(\boldsymbol{y}_1^*, \boldsymbol{y}_2^*, \cdots, \boldsymbol{y}_m^*)$ must be a solution of (2.13). If not, then the system of equations (2.13) is inconsistent. In other words, there is no Nash equilibrium of followers in the given bilevel programming. Although this method can deal with general problem, it is a slow way to find a Nash equilibrium.

After obtaining the Nash equilibrium for each given decision vector of the leader, we may compute the objective value of the leader for each given control vector according to the Nash equilibrium. Hence we may employ the GA to search for the Stackelberg-Nash equilibrium.

Now we consider a bilevel programming with three followers in which the leader has a decision vector (x_1, x_2, x_3) and the three followers have decision vector $(y_{11}, y_{12}, y_{21}, y_{22}, y_{31}, y_{32})$,

$$
\left\{
\begin{array}{l}
\max\limits_{x_1,x_2,x_3} y_{11}^* y_{12}^* \sin x_1 + 2y_{21}^* y_{22}^* \sin x_2 + 3y_{31}^* y_{32}^* \sin x_3 \\
\text{subject to:} \\
\qquad x_1 + x_2 + x_3 \le 10,\ x_1 \ge 0,\ x_2 \ge 0,\ x_3 \ge 0 \\
\qquad (y_{11}^*, y_{12}^*, y_{21}^*, y_{22}^*, y_{31}^*, y_{32}^*) \text{ solves the problems} \\
\qquad \left\{
\begin{array}{l}
\max\limits_{y_{11},y_{12}} y_{11} \sin y_{12} + y_{12} \sin y_{11} \\
\text{subject to:} \\
\qquad y_{11} + y_{12} \le x_1,\ y_{11} \ge 0,\ y_{12} \ge 0
\end{array}
\right. \\
\qquad \left\{
\begin{array}{l}
\max\limits_{y_{21},y_{22}} y_{21} \sin y_{22} + y_{22} \sin y_{21} \\
\text{subject to:} \\
\qquad y_{21} + y_{22} \le x_2,\ y_{21} \ge 0,\ y_{22} \ge 0
\end{array}
\right. \\
\qquad \left\{
\begin{array}{l}
\max\limits_{y_{31},y_{32}} y_{31} \sin y_{32} + y_{32} \sin y_{31} \\
\text{subject to:} \\
\qquad y_{31} + y_{32} \le x_3,\ y_{31} \ge 0,\ y_{32} \ge 0.
\end{array}
\right.
\end{array}
\right.
$$

A run of GA with 1000 generations shows that the Stackelberg-Nash equilibrium is

$$
\begin{aligned}
(x_1^*, x_2^*, x_3^*) &= (0.000, 1.936, 8.064), \\
(y_{11}^*, y_{12}^*) &= (0.000, 0.000), \\
(y_{21}^*, y_{22}^*) &= (0.968, 0.968), \\
(y_{31}^*, y_{32}^*) &= (1.317, 6.747)
\end{aligned}
$$

with optimal objective values

$$
\begin{aligned}
y_{11}^* y_{12}^* \sin x_1^* + 2y_{21}^* y_{22}^* \sin x_2^* + 3y_{31}^* y_{32}^* \sin x_3^* &= 27.822, \\
y_{11}^* \sin y_{12}^* + y_{12}^* \sin y_{11}^* &= 0.000, \\
y_{21}^* \sin y_{22}^* + y_{22}^* \sin y_{21}^* &= 1.595, \\
y_{31}^* \sin y_{32}^* + y_{32}^* \sin y_{31}^* &= 7.120.
\end{aligned}
$$

Chapter 3
Neural Networks

Neural network (NN), inspired by the current understanding of biological NN, is a class of adaptive systems consisting of a number of simple processing elements, called neurons, that are interconnected to each other in a feedforward way. Although NN can perform some human brain-like tasks, there is still a huge gap between biological and artificial NN. An important contribution of NN is the ability to learn to perform operations, not only for inputs exactly like the training data, but also for new data that may be incomplete or noisy. NN has also the benefit of easy modification by retraining with an updated data set. For our purpose, the significant advantage of NN is the speed of operation after it is trained.

3.1 Basic Concepts

The artificial neuron simulates the behavior of the biological neuron to make a simple operation of a weighted sum of the incoming signals as

$$y = w_0 + w_1x_1 + w_2x_2 + \cdots + w_nx_n \tag{3.1}$$

where $x_1, x_2, \cdots, x_n$ are inputs, $w_0, w_1, w_2, \cdots, w_n$ are weights, and y is output. Figure 3.1 illustrates a neuron.

In most applications, we define a memoryless nonlinear function σ as an activation function to change the output to

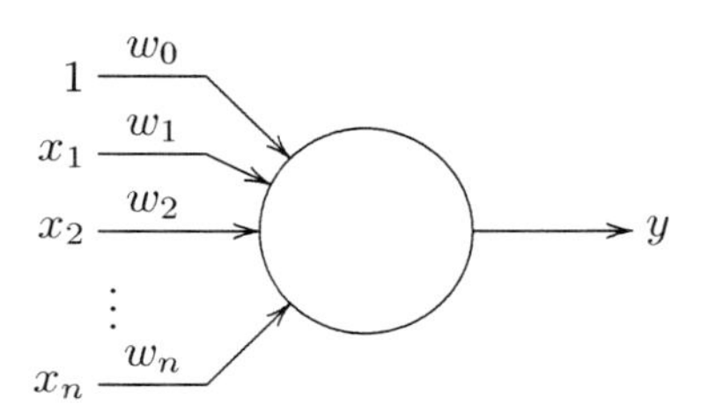

Fig. 3.1 An Artificial Neuron

B. Liu: Theory and Practice of Uncertain Programming, STUDFUZZ 239, pp. 19–24.
springerlink.com

$$y = \sigma\left(w_0 + w_1 x_1 + w_2 x_2 + \cdots + w_n x_n\right). \tag{3.2}$$

The choice of the activation functions depends on the application area. In this book we employ the sigmoid function defined as

$$\sigma(x) = \frac{1}{1 + e^{-x}} \tag{3.3}$$

whose derivative is

$$\sigma'(x) = \frac{e^{-x}}{(1 + e^{-x})^2}. \tag{3.4}$$

They are shown in Figure 3.2.

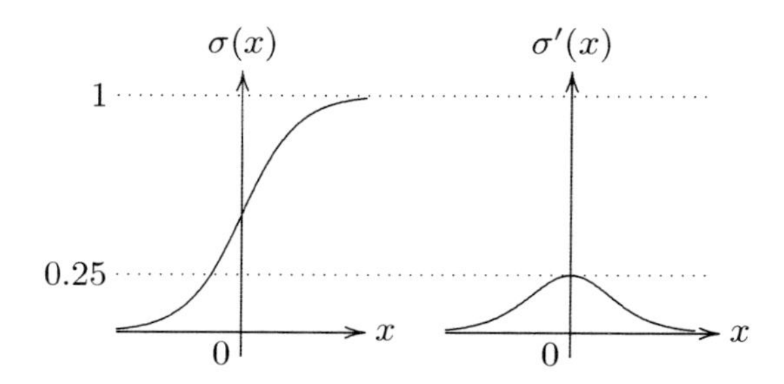

Fig. 3.2 Sigmoid Function and Derivative

Let us consider an NN with one hidden layer, in which there are n neurons in the input layer, m neurons in the output layer, and p neurons in the hidden layer which is pictured in Figure 3.3. Then the outputs of the neurons in the hidden layer are

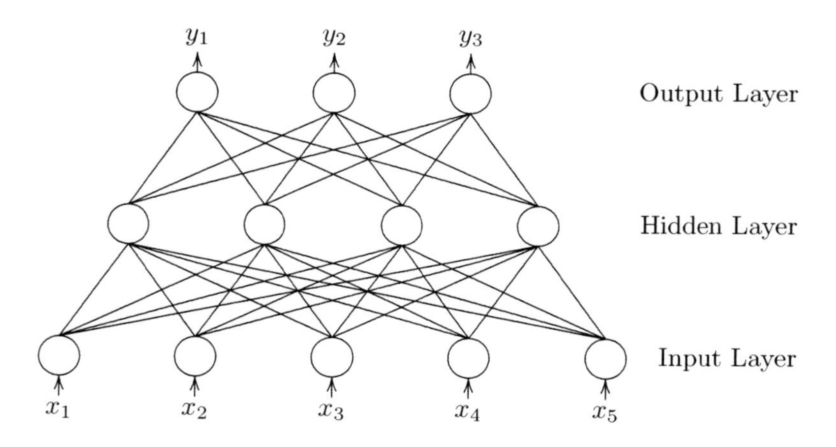

Fig. 3.3 A Neural Network

$$x_i^1 = \sigma\left(\sum_{j=1}^{n} w_{ij}^0 x_j + w_{i0}^0\right), \quad i = 1, 2, \cdots, p. \tag{3.5}$$

Thus the outputs of the neurons in the output layer are

$$y_i = \sum_{j=1}^{p} w_{ij}^1 x_j^1 + w_{i0}^1, \quad i = 1, 2, \cdots, m. \tag{3.6}$$

The coefficients $w_{ij}^0, i = 1, 2, \cdots, p, j = 0, 1, \cdots, n$ in (3.5) and $w_{ij}^1, i = 1, 2, \cdots, m, j = 0, 1, \cdots, p$ in (3.6) are called the network weights.

3.2 Function Approximation

NN is clearly a nonlinear mapping from the input space to the output space. It has been proved that any continuous nonlinear function can be approximated arbitrarily well over a compact set by an NN consisting of one hidden layer provided that there are sufficiently many neurons in the hidden layer.

Assume that $f : \Re^n \to \Re^m$ is a continuous nonlinear function. We hope to train an NN to approximate the function $f(\boldsymbol{x})$. For an NN with a fixed number of neurons and architecture, the network weights may be arranged into a vector $\boldsymbol{w}$. Let $F(\boldsymbol{x}, \boldsymbol{w})$ be the output of mapping implemented by the NN.

The training process is to find an appropriate weight vector $\boldsymbol{w}$ that provides the best possible approximation of $f(\boldsymbol{x})$. Let

$$(\boldsymbol{x}_1^*, \boldsymbol{y}_1^*), \quad (\boldsymbol{x}_2^*, \boldsymbol{y}_2^*), \quad \cdots, \quad (\boldsymbol{x}_N^*, \boldsymbol{y}_N^*)$$

be a set of training input-output data on $f(\boldsymbol{x})$. We wish to choose a weight vector $\boldsymbol{w}$ so that the output $F(\boldsymbol{x}, \boldsymbol{w})$ is "close" to the desired output $\boldsymbol{y}_i^*$ for the input $\boldsymbol{x}_i^*$. That is, the training process is to find the weight vector $\boldsymbol{w}$ that minimizes the following error function,

$$Err(\boldsymbol{w}) = \frac{1}{2}\sum_{i=1}^{N} \|F(\boldsymbol{x}_i^*, \boldsymbol{w}) - \boldsymbol{y}_i^*\|^2. \tag{3.7}$$

3.3 Neuron Number Determination

Since the function f is a mapping from $\Re^n$ to $\Re^m$, the number of input neurons is always n, and the number of output neurons is always m. Thus the main problem is to determine the best number of hidden neurons.

Although any continuous function can be approximated with an arbitrary accuracy by an NN with infinite neurons in the hidden layer, it is practically impossible to have infinite hidden neurons. On the one hand, too few hidden neurons make the NN lack of generalization ability. On the other hand, too many hidden neurons increase the training time and response time of the trained NN. A lot of methods have been proposed for determining the number of hidden neurons, some add hidden neurons, and other delete hidden neurons during the training process.

3.4 Backpropagation Algorithm

The values of the weights represent all memory of the NN. During the training phase of an NN, the values of weights are continuously updated by the training process until some termination criterion is met. In other words, learning in NN is a modification process of the values of weights so as to bring the mapping implemented by the NN as close as possible to a desired mapping. It may also be viewed as an optimization problem of selecting weights to minimize the error between the target output and the actual output.

Backpropagation algorithm is an effective learning algorithm. It is essentially a gradient method. Here we introduce the backpropagation algorithm for the NN with one hidden layer. Assume that there are N samples $(x_{k1}^*, x_{k2}^*, \cdots, x_{kn}^*; y_{k1}^*, y_{k2}^*, \cdots, y_{km}^*)$ for $k = 1, 2, \cdots, N$.

We first initialize the weight vector $\boldsymbol{w}$ at random, set $\Delta w_{ij}^1 = 0$ for $i = 1, 2, \cdots, m$, $j = 0, 1, \cdots, p$, $\Delta w_{ij}^0 = 0$ for $i = 1, 2, \cdots, p$ and $j = 0, 1, \cdots, n$, and the adaptive parameter $\lambda = 1$. Then we adjust the weights by an on-line learning process. When the k-th sample is used, the outputs of the hidden neurons are

$$x_{ki}^1 = \sigma\left(\sum_{j=1}^{n} w_{ij}^0 x_{kj}^* + w_{i0}^0\right), \quad i = 1, 2, \cdots, p,$$

and the outputs of the NN are

$$y_{ki} = \sum_{j=1}^{p} w_{ij}^1 x_{kj}^1 + w_{i0}^1, \quad i = 1, 2, \cdots, m.$$

In order to speed up the learning process, we use an improved error function

$$E_k = \frac{1}{2}\sum_{i=1}^{m} \left[\lambda(y_{ki}^* - y_{ki})^2 + (1-\lambda)\Phi(y_{ki}^* - y_{ki})\right] \tag{3.8}$$

where $\Phi(x) = \ln(\cosh(\beta x))/\beta$ and β is a constant, for example, $\beta = 4/3$.

Thus the equations for weight change are given as follows: For the hidden-output weights $w_{ij}^1, i = 1, 2, \cdots, m, j = 0, 1, \cdots, p$, we have

$$\Delta w_{ij}^1 \leftarrow -\alpha\frac{\partial E_k}{\partial w_{ij}^1} + \eta\Delta w_{ij}^1 = \alpha C_i^1 x_{kj}^1 + \eta\Delta w_{ij}^1 \tag{3.9}$$

where

$$C_i^1 = \lambda(y_{ki}^* - y_{ki}) + (1-\lambda)\tanh(\beta(y_{ki}^* - y_{ki})), \quad x_{k0}^1 = 1.$$

For the input-hidden weights $w_{ij}^0, i = 1, 2, \cdots, p, j = 0, 1, \cdots, n$, we have

$$\Delta w_{ij}^0 \leftarrow -\alpha \frac{\partial E_k}{\partial w_{ij}^0} + \eta \Delta w_{ij}^0 = \alpha C_i^0 x_{kj}^* + \eta \Delta w_{ij}^0 \tag{3.10}$$

where

$$C_i^0 = \left(1 - (x_{ki}^1)^2\right) \sum_{l=1}^{m} C_l^1 w_{li}^1, \quad x_{k0}^* = 1,$$

α and η are numbers between 0 and 1, for example, $\alpha = 0.05$, $\eta = 0.01$.

After training the NN one time over all input-output data, we calculate the total error $E = E_1 + E_2 + \cdots + E_N$. If the error E is less than a predetermined precision E_0 (for example, 0.05), then the NN is considered trained well. Otherwise, we set $\lambda = \exp(-1/E^2)$ and repeat the learning process until $E < E_0$.

Algorithm 3.1 (Backpropagation Algorithm)
Step 1. Initialize weight vector $\boldsymbol{w}$, and set $\lambda = 1$ and $k = 0$.
Step 2. $k \leftarrow k + 1$.
Step 3. Adjust the weight vector $\boldsymbol{w}$ according to (3.9) and (3.10).
Step 4. Calculate the error E_k according to (3.8).
Step 5. If $k < N$, go to Step 2.
Step 6. Set $E = E_1 + E_2 + \cdots + E_N$.
Step 7. If $E > E_0$, then $k = 0$, $\lambda = \exp(-1/E^2)$ and go to Step 2.
Step 8. End.

3.5 Numerical Experiments

The NN architecture used in this section is the network with one hidden layer which is pictured in Figure 3.3. The NN will be used for approximating some continuous functions.

Example 3.1. Let us design an NN to approximate the continuous function,

$$f(x_1, x_2, x_3, x_4) = \sin x_1 + \sin x_2 + \sin x_3 + \sin x_4$$

defined on $[0, 2\pi]^4$. In order to approximate the function $f(x_1, x_2, x_3, x_4)$ by an NN, we generate 3000 input-output data. Then we train an NN (4 input neurons, 10 hidden neurons, 1 output neuron) by the backpropagation algorithm. The sum-squared error of the trained NN is 0.51, and the average error is 0.01.

Example 3.2. Consider the vector-valued continuous function,

$$f(x_1, x_2, x_3, x_4, x_5, x_6) = \begin{pmatrix} x_1 \ln x_2 + x_2 \ln x_3 \\ x_3 \ln x_4 + x_4 \ln x_5 \\ x_5 \ln x_6 + x_6 \ln x_1 \end{pmatrix}$$

defined on the region $[1, 5]^6$. We generate 3000 training data for the function $f(\boldsymbol{x})$. Then we train an NN (6 input neurons, 15 hidden neurons, 3 output neurons) to approximate the function. A run of backpropagation algorithm shows that the sum-squared error of the trained NN is 6.16, and the average error is 0.05.

Example 3.3. Consider the function

$$f(x_1, x_2, x_3, x_4) = \frac{x_1}{1+x_1} + \frac{x_2}{1+x_2} + \frac{x_3}{1+x_3} + \frac{x_4}{1+x_4}$$

defined on $[0, 2]$. Assume that the input-output data for the function $f(\boldsymbol{x})$ are randomly generated on $[0, 2]$ with a uniformly distributed noise $\mathcal{U}(-a, a)$, where $\mathcal{U}(-a, a)$ represents the uniformly distributed variable on the interval $[-a, a]$. That is, for each input $\boldsymbol{x}$, the output $y = f(\boldsymbol{x}) + \mathcal{U}(-a, a)$.

We produce 2000 input-output data $\{(\boldsymbol{x}_i^*, y_i^*) | i = 1, 2, \cdots, 2000\}$ with noise $\mathcal{U}(-a, a)$ for the function $f(\boldsymbol{x})$. The backpropagation algorithm may obtain an NN (4 input neurons, 6 hidden neurons, 1 output neuron) to approximate the function $f(\boldsymbol{x})$ according to the noise data.

Let $F(\boldsymbol{x}, \boldsymbol{w}^*)$ be the output of mapping implemented by the NN. We generate 1000 test noise data $\{(\boldsymbol{x}_i^*, y_i^*) | i = 2001, 2002, \cdots, 3000\}$, then we have the errors shown in Table 3.1.

Table 3.1 Sum-Squared Errors

Noise	$\frac{1}{2}\sum_{i=2001}^{3000} \lvert F(\boldsymbol{x}_i^*, \boldsymbol{w}^*) - f(\boldsymbol{x}_i^*)\rvert^2$	$\frac{1}{2}\sum_{i=2001}^{3000} \lvert y_i^* - f(\boldsymbol{x}_i^*)\rvert^2$
$\mathcal{U}(-0.05, 0.05)$	0.362	0.389
$\mathcal{U}(-0.10, 0.10)$	1.333	1.643
$\mathcal{U}(-0.20, 0.20)$	4.208	6.226
$\mathcal{U}(-0.30, 0.30)$	7.306	14.01
$\mathcal{U}(-0.40, 0.40)$	14.74	24.90

Note that the errors in the first column are less than that in the second column. This means that the trained NN can compensate for the error of the noise training data.

Chapter 4
Stochastic Programming

With the requirement of considering randomness, different types of stochastic programming have been developed to suit the different purposes of management. The first type of stochastic programming is the *expected value model*, which optimizes the expected objective functions subject to some expected constraints. The second, *chance-constrained programming*, was pioneered by Charnes and Cooper [37] as a means of handling uncertainty by specifying a confidence level at which it is desired that the stochastic constraint holds. After that, Liu [174] generalized chance-constrained programming to the case with not only stochastic constraints but also stochastic objectives. In practice, there usually exist multiple events in a complex stochastic decision system. Sometimes the decision-maker wishes to maximize the chance functions of satisfying these events. In order to model this type of problem, Liu [166] provided a theoretical framework of the third type of stochastic programming, called *dependent-chance programming*.

This chapter will give some basic concepts of probability theory and introduce a spectrum of stochastic programming. A hybrid intelligent algorithm is also documented.

4.1 Random Variables

Before introducing the concept of random variable, let us define a probability measure by an axiomatic approach.

Definition 4.1. *Let Ω be a nonempty set, and $\mathcal{A}$ a σ-algebra of subsets (called events) of Ω. The set function* Pr *is called a probability measure if*
Axiom 1. *(Normality)* $\Pr\{\Omega\} = 1$*;*
Axiom 2. *(Nonnegativity)* $\Pr\{A\} \geq 0$ *for any event A;*
Axiom 3. *(Countable Additivity) For every countable sequence of mutually disjoint events $\{A_i\}$, we have*

B. Liu: Theory and Practice of Uncertain Programming, STUDFUZZ 239, pp. 25–56.
springerlink.com

$$\Pr\left\{\bigcup_{i=1}^{\infty} A_i\right\} = \sum_{i=1}^{\infty} \Pr\{A_i\}. \tag{4.1}$$

Example 4.1. Let $\Omega = \{\omega_1, \omega_2, \cdots\}$, and let $\mathcal{A}$ be the power set of Ω. Assume that $p_1, p_2, \cdots$ are nonnegative numbers such that $p_1 + p_2 + \cdots = 1$. Define a set function on $\mathcal{A}$ as

$$\Pr\{A\} = \sum_{\omega_i \in A} p_i, \quad A \in \mathcal{A}. \tag{4.2}$$

Then Pr is a probability measure.

Example 4.2. Let $\Omega = [0, 1]$ and let $\mathcal{A}$ be the Borel algebra over Ω. If Pr is the Lebesgue measure, then Pr is a probability measure.

Example 4.3. Let ϕ be a nonnegative and integrable function on $\Re$ (the set of real numbers) such that $\int_{\Re} \phi(x)\mathrm{d}x = 1$. Then for any Borel set A, the set function

$$\Pr\{A\} = \int_A \phi(x)\mathrm{d}x \tag{4.3}$$

is a probability measure on $\Re$.

Theorem 4.1. *Let Ω be a nonempty set, $\mathcal{A}$ a σ-algebra over Ω, and* Pr *a probability measure. Then* $\Pr\{\emptyset\} = 0$ *and* $0 \le \Pr\{A\} \le 1$ *for any event A.*

Proof: Since $\emptyset$ and Ω are disjoint events and $\emptyset \cup \Omega = \Omega$, we have $\Pr\{\emptyset\} + \Pr\{\Omega\} = \Pr\{\Omega\}$ which makes $\Pr\{\emptyset\} = 0$. By the nonnegativity axiom, we have $\Pr\{A\} \ge 0$ for any event A. By the countable additivity axiom, we get $\Pr\{A\} = 1 - \Pr\{A^c\} \le 1$.

Definition 4.2. *Let Ω be a nonempty set, $\mathcal{A}$ a σ-algebra of subsets of Ω, and* Pr *a probability measure. Then the triplet* $(\Omega, \mathcal{A}, \Pr)$ *is called a probability space.*

Definition 4.3. *A random variable is a measurable function from a probability space* $(\Omega, \mathcal{A}, \Pr)$ *to the set of real numbers, i.e., for any Borel set B of real numbers, the set*

$$\{\xi \in B\} = \{\omega \in \Omega \mid \xi(\omega) \in B\} \tag{4.4}$$

is an event.

Definition 4.4. *Let $f : \Re^n \to \Re$ be a measurable function, and $\xi_1, \xi_2, \cdots, \xi_n$ random variables defined on the probability space* $(\Omega, \mathcal{A}, \Pr)$. *Then $\xi = f(\xi_1, \xi_2, \cdots, \xi_n)$ is a random variable defined by*

$$\xi(\omega) = f(\xi_1(\omega), \xi_2(\omega), \cdots, \xi_n(\omega)), \quad \forall \omega \in \Omega. \tag{4.5}$$

Definition 4.5. *An n-dimensional random vector is a measurable function from a probability space $(\Omega, \mathcal{A}, \Pr)$ to the set of n-dimensional real vectors, i.e., for any Borel set B of $\Re^n$, the set*

$$\{\boldsymbol{\xi} \in B\} = \{\omega \in \Omega \mid \boldsymbol{\xi}(\omega) \in B\} \tag{4.6}$$

is an event.

Theorem 4.2. *The vector $(\xi_1, \xi_2, \cdots, \xi_n)$ is a random vector if and only if $\xi_1, \xi_2, \cdots, \xi_n$ are random variables.*

Proof: Write $\boldsymbol{\xi} = (\xi_1, \xi_2, \cdots, \xi_n)$. Suppose that $\boldsymbol{\xi}$ is a random vector on the probability space $(\Omega, \mathcal{A}, \Pr)$. For any Borel set B of $\Re$, the set $B \times \Re^{n-1}$ is also a Borel set of $\Re^n$. Thus we have

$$\begin{aligned}\{\omega \in \Omega \mid \xi_1(\omega) \in B\} &= \{\omega \in \Omega \mid \xi_1(\omega) \in B, \xi_2(\omega) \in \Re, \cdots, \xi_n(\omega) \in \Re\} \\ &= \{\omega \in \Omega \mid \boldsymbol{\xi}(\omega) \in B \times \Re^{n-1}\} \in \mathcal{A}\end{aligned}$$

which implies that ξ_1 is a random variable. A similar process may prove that $\xi_2, \xi_3, \cdots, \xi_n$ are random variables. Conversely, suppose that all $\xi_1, \xi_2, \cdots, \xi_n$ are random variables on the probability space $(\Omega, \mathcal{A}, \Pr)$. We define

$$\mathcal{B} = \left\{B \subset \Re^n \mid \{\omega \in \Omega | \boldsymbol{\xi}(\omega) \in B\} \in \mathcal{A}\right\}.$$

The vector $\boldsymbol{\xi} = (\xi_1, \xi_2, \cdots, \xi_n)$ is proved to be a random vector if we can prove that $\mathcal{B}$ contains all Borel sets of $\Re^n$. First, the class $\mathcal{B}$ contains all open intervals of $\Re^n$ because

$$\left\{\omega \in \Omega \mid \boldsymbol{\xi}(\omega) \in \prod_{i=1}^n (a_i, b_i)\right\} = \bigcap_{i=1}^n \{\omega \in \Omega \mid \xi_i(\omega) \in (a_i, b_i)\} \in \mathcal{A}.$$

Next, the class $\mathcal{B}$ is a σ-algebra of $\Re^n$ because (i) we have $\Re^n \in \mathcal{B}$ since $\{\omega \in \Omega | \boldsymbol{\xi}(\omega) \in \Re^n\} = \Omega \in \mathcal{A}$; (ii) if $B \in \mathcal{B}$, then $\{\omega \in \Omega | \boldsymbol{\xi}(\omega) \in B\} \in \mathcal{A}$, and

$$\{\omega \in \Omega \mid \boldsymbol{\xi}(\omega) \in B^c\} = \{\omega \in \Omega \mid \boldsymbol{\xi}(\omega) \in B\}^c \in \mathcal{A}$$

which implies that $B^c \in \mathcal{B}$; (iii) if $B_i \in \mathcal{B}$ for $i = 1, 2, \cdots$, then $\{\omega \in \Omega | \boldsymbol{\xi}(\omega) \in B_i\} \in \mathcal{A}$ and

$$\left\{\omega \in \Omega \mid \boldsymbol{\xi}(\omega) \in \bigcup_{i=1}^\infty B_i\right\} = \bigcup_{i=1}^\infty \{\omega \in \Omega \mid \boldsymbol{\xi}(\omega) \in B_i\} \in \mathcal{A}$$

which implies that $\cup_i B_i \in \mathcal{B}$. Since the smallest σ-algebra containing all open intervals of $\Re^n$ is just the Borel algebra of $\Re^n$, the class $\mathcal{B}$ contains all Borel sets of $\Re^n$. The theorem is proved.

Probability Distribution

Definition 4.6. *The probability distribution Φ: $\Re \to [0,1]$ of a random variable ξ is defined by*

$$\Phi(x) = \Pr\left\{\omega \in \Omega \;\middle|\; \xi(\omega) \le x\right\}. \tag{4.7}$$

That is, $\Phi(x)$ is the probability that the random variable ξ takes a value less than or equal to x.

Definition 4.7. *The probability density function ϕ: $\Re \to [0,+\infty)$ of a random variable ξ is a function such that*

$$\Phi(x) = \int_{-\infty}^{x} \phi(y)\mathrm{d}y \tag{4.8}$$

holds for all $x \in \Re$, where Φ is the probability distribution of the random variable ξ.

Uniform Distribution: A random variable ξ has a uniform distribution if its probability density function is

$$\phi(x) = \begin{cases} \dfrac{1}{b-a}, & \text{if } a \le x \le b \\ 0, & \text{otherwise} \end{cases} \tag{4.9}$$

denoted by $\mathcal{U}(a,b)$, where a and b are given real numbers with $a < b$.

Exponential Distribution: A random variable ξ has an exponential distribution if its probability density function is

$$\phi(x) = \begin{cases} \dfrac{1}{\beta}\exp\left(-\dfrac{x}{\beta}\right), & \text{if } x \ge 0 \\ 0, & \text{otherwise} \end{cases} \tag{4.10}$$

denoted by $\mathcal{EXP}(\beta)$, where β is a positive number.

Normal Distribution: A random variable ξ has a normal distribution if its probability density function is

$$\phi(x) = \frac{1}{\sigma\sqrt{2\pi}}\exp\left(-\frac{(x-\mu)^2}{2\sigma^2}\right), \quad x \in \Re \tag{4.11}$$

denoted by $\mathcal{N}(\mu,\sigma^2)$, where μ and σ are real numbers.

Theorem 4.3 *(Probability Inversion Theorem). Let ξ be a random variable whose probability density function ϕ exists. Then for any Borel set B of $\Re$, we have*

$$\Pr\{\xi \in B\} = \int_{B} \phi(y)\mathrm{d}y. \tag{4.12}$$

Proof: Let $\mathcal{C}$ be the class of all subsets C of $\Re$ for which the relation

$$\Pr\{\xi \in C\} = \int_C \phi(y)\mathrm{d}y \tag{4.13}$$

holds. We will show that $\mathcal{C}$ contains all Borel sets of $\Re$. It follows from the probability continuity theorem and relation (4.13) that $\mathcal{C}$ is a monotone class. It is also clear that $\mathcal{C}$ contains all intervals of the form $(-\infty, a]$, $(a, b]$, (b, ∞) and $\emptyset$ since

$$\Pr\{\xi \in (-\infty, a]\} = \Phi(a) = \int_{-\infty}^{a} \phi(y)\mathrm{d}y,$$

$$\Pr\{\xi \in (b, +\infty)\} = \Phi(+\infty) - \Phi(b) = \int_{b}^{+\infty} \phi(y)\mathrm{d}y,$$

$$\Pr\{\xi \in (a, b]\} = \Phi(b) - \Phi(a) = \int_{a}^{b} \phi(y)\mathrm{d}y,$$

$$\Pr\{\xi \in \emptyset\} = 0 = \int_{\emptyset} \phi(y)\mathrm{d}y$$

where Φ is the probability distribution of ξ. Let $\mathcal{F}$ be the algebra consisting of all finite unions of disjoint sets of the form $(-\infty, a]$, $(a, b]$, (b, ∞) and $\emptyset$. Note that for any disjoint sets $C_1, C_2, \cdots, C_m$ of $\mathcal{F}$ and $C = C_1 \cup C_2 \cup \cdots \cup C_m$, we have

$$\Pr\{\xi \in C\} = \sum_{j=1}^{m} \Pr\{\xi \in C_j\} = \sum_{j=1}^{m} \int_{C_j} \phi(y)\mathrm{d}y = \int_C \phi(y)\mathrm{d}y.$$

That is, $C \in \mathcal{C}$. Hence we have $\mathcal{F} \subset \mathcal{C}$. Since the smallest σ-algebra containing $\mathcal{F}$ is just the Borel algebra of $\Re$, the monotone class theorem implies that $\mathcal{C}$ contains all Borel sets of $\Re$.

Example 4.4. Let ξ be a uniformly distributed random variable on $[a, b]$. Then for any number $c \in [a, b]$, it follows from probability inversion theorem that

$$\Pr\{\xi \le c\} = \int_a^c \phi(x)\mathrm{d}x = \int_a^c \frac{1}{b-a}\mathrm{d}x = \frac{c-a}{b-a}.$$

Independence

Definition 4.8. *The random variables $\xi_1, \xi_2, \cdots, \xi_m$ are said to be independent if*

$$\Pr\left\{\bigcap_{i=1}^{m}\{\xi_i \in B_i\}\right\} = \prod_{i=1}^{m} \Pr\{\xi_i \in B_i\} \tag{4.14}$$

for any Borel sets $B_1, B_2, \cdots, B_m$ of real numbers.

Theorem 4.4. *Let ξ_i be random variables with probability distributions Φ_i, $i = 1, 2, \cdots, m$, respectively, and Φ the probability distribution of the random vector $(\xi_1, \xi_2, \cdots, \xi_m)$. Then $\xi_1, \xi_2, \cdots, \xi_m$ are independent if and only if*

$$\Phi(x_1, x_2, \cdots, x_m) = \Phi_1(x_1)\Phi_2(x_2)\cdots\Phi_m(x_m) \tag{4.15}$$

for all $(x_1, x_2, \cdots, x_m) \in \Re^m$.

Proof: If $\xi_1, \xi_2, \cdots, \xi_m$ are independent random variables, then we have

$$\begin{aligned}\Phi(x_1, x_2, \cdots, x_m) &= \Pr\{\xi_1 \le x_1, \xi_2 \le x_2, \cdots, \xi_m \le x_m\} \\ &= \Pr\{\xi_1 \le x_1\}\Pr\{\xi_2 \le x_2\}\cdots\Pr\{\xi_m \le x_m\} \\ &= \Phi_1(x_1)\Phi_2(x_2)\cdots\Phi_m(x_m)\end{aligned}$$

for all $(x_1, x_2, \cdots, x_m) \in \Re^m$. Conversely, assume that (4.15) holds. Let $x_2, x_3, \cdots, x_m$ be fixed real numbers, and $\mathfrak{C}$ the class of all subsets C of $\Re$ for which the relation

$$\Pr\{\xi_1 \in C, \xi_2 \le x_2, \cdots, \xi_m \le x_m\} = \Pr\{\xi_1 \in C\}\prod_{i=2}^{m}\Pr\{\xi_i \le x_i\} \tag{4.16}$$

holds. We will show that $\mathfrak{C}$ contains all Borel sets of $\Re$. It follows from the probability continuity theorem and relation (4.16) that $\mathfrak{C}$ is a monotone class. It is also clear that $\mathfrak{C}$ contains all intervals of the form $(-\infty, a]$, $(a, b]$, (b, ∞) and $\emptyset$. Let $\mathfrak{F}$ be the algebra consisting of all finite unions of disjoint sets of the form $(-\infty, a]$, $(a, b]$, (b, ∞) and $\emptyset$. Note that for any disjoint sets $C_1, C_2, \cdots, C_k$ of $\mathfrak{F}$ and $C = C_1 \cup C_2 \cup \cdots \cup C_k$, we have

$$\begin{aligned}&\Pr\{\xi_1 \in C, \xi_2 \le x_2, \cdots, \xi_m \le x_m\} \\ &= \sum_{j=1}^{m}\Pr\{\xi_1 \in C_j, \xi_2 \le x_2, \cdots, \xi_m \le x_m\} \\ &= \Pr\{\xi_1 \in C\}\Pr\{\xi_2 \le x_2\}\cdots\Pr\{\xi_m \le x_m\}.\end{aligned}$$

That is, $C \in \mathfrak{C}$. Hence we have $\mathfrak{F} \subset \mathfrak{C}$. Since the smallest σ-algebra containing $\mathfrak{F}$ is just the Borel algebra of $\Re$, the monotone class theorem implies that $\mathfrak{C}$ contains all Borel sets of $\Re$. Applying the same reasoning to each ξ_i in turn, we obtain the independence of the random variables.

Theorem 4.5. *Let ξ_i be random variables with probability density functions ϕ_i, $i = 1, 2, \cdots, m$, respectively, and ϕ the probability density function of the random vector $(\xi_1, \xi_2, \cdots, \xi_m)$. Then $\xi_1, \xi_2, \cdots, \xi_m$ are independent if and only if*

$$\phi(x_1, x_2, \cdots, x_m) = \phi_1(x_1)\phi_2(x_2)\cdots\phi_m(x_m) \tag{4.17}$$

for almost all $(x_1, x_2, \cdots, x_m) \in \Re^m$.

Proof: If $\phi(x_1, x_2, \cdots, x_m) = \phi_1(x_1)\phi_2(x_2)\cdots\phi_m(x_m)$ a.e., then we have

$$\begin{aligned}
\Phi(x_1, x_2, \cdots, x_m) &= \int_{-\infty}^{x_1}\int_{-\infty}^{x_2}\cdots\int_{-\infty}^{x_m}\phi(t_1, t_2, \cdots, t_m)\mathrm{d}t_1\mathrm{d}t_2\cdots\mathrm{d}t_m \\
&= \int_{-\infty}^{x_1}\int_{-\infty}^{x_2}\cdots\int_{-\infty}^{x_m}\phi_1(t_1)\phi_2(t_2)\cdots\phi_m(t_m)\mathrm{d}t_1\mathrm{d}t_2\cdots\mathrm{d}t_m \\
&= \int_{-\infty}^{x_1}\phi_1(t_1)\mathrm{d}t_1\int_{-\infty}^{x_2}\phi_2(t_2)\mathrm{d}t_2\cdots\int_{-\infty}^{x_m}\phi_m(t_m)\mathrm{d}t_m \\
&= \Phi_1(x_1)\Phi_2(x_2)\cdots\Phi_m(x_m)
\end{aligned}$$

for all $(x_1, x_2, \cdots, x_m) \in \Re^m$. Thus $\xi_1, \xi_2, \cdots, \xi_m$ are independent. Conversely, if $\xi_1, \xi_2, \cdots, \xi_m$ are independent, then for any $(x_1, x_2, \cdots, x_m) \in \Re^m$, we have $\Phi(x_1, x_2, \cdots, x_m) = \Phi_1(x_1)\Phi_2(x_2)\cdots\Phi_m(x_m)$. Hence

$$\Phi(x_1, x_2, \cdots, x_m) = \int_{-\infty}^{x_1}\int_{-\infty}^{x_2}\cdots\int_{-\infty}^{x_m}\phi_1(t_1)\phi_2(t_2)\cdots\phi_m(t_m)\mathrm{d}t_1\mathrm{d}t_2\cdots\mathrm{d}t_m$$

which implies that $\phi(x_1, x_2, \cdots, x_m) = \phi_1(x_1)\phi_2(x_2)\cdots\phi_m(x_m)$ a.e.

Example 4.5. Let $\xi_1, \xi_2, \cdots, \xi_m$ be independent random variables with probability density functions $\phi_1, \phi_2, \cdots, \phi_m$, respectively, and $f : \Re^m \to \Re$ a measurable function. Then for any Borel set B of real numbers, the probability $\Pr\{f(\xi_1, \xi_2, \cdots, \xi_m) \in B\}$ is

$$\iint\cdots\int_{f(x_1, x_2, \cdots, x_m)\in B}\phi_1(x_1)\phi_2(x_2)\cdots\phi_m(x_m)\mathrm{d}x_1\mathrm{d}x_2\cdots\mathrm{d}x_m.$$

Expected Value

Expected value is the average value of random variable in the sense of probability measure. It may be defined as follows.

Definition 4.9. *Let ξ be a random variable. Then the expected value of ξ is defined by*

$$E[\xi] = \int_0^{+\infty}\Pr\{\xi \ge r\}\mathrm{d}r - \int_{-\infty}^{0}\Pr\{\xi \le r\}\mathrm{d}r \tag{4.18}$$

provided that at least one of the two integrals is finite.

Let ξ and η be random variables with finite expected values. For any numbers a and b, it has been proved that $E[a\xi + b\eta] = aE[\xi] + bE[\eta]$. That is, the expected value operator has the linearity property.

Example 4.6. Assume that ξ is a discrete random variable taking values x_i with probabilities p_i, $i = 1, 2, \cdots, m$, respectively. It follows from the definition of expected value operator that

$$E[\xi] = \sum_{i=1}^{m} p_i x_i.$$

Theorem 4.6. *Let ξ be a random variable whose probability density function ϕ exists. If the Lebesgue integral*

$$\int_{-\infty}^{+\infty} x\phi(x)\mathrm{d}x$$

is finite, then we have

$$E[\xi] = \int_{-\infty}^{+\infty} x\phi(x)\mathrm{d}x. \tag{4.19}$$

Proof: It follows from Definition 4.9 and Fubini Theorem that

$$\begin{aligned}
E[\xi] &= \int_0^{+\infty} \Pr\{\xi \geq r\}\mathrm{d}r - \int_{-\infty}^{0} \Pr\{\xi \leq r\}\mathrm{d}r \\
&= \int_0^{+\infty} \left[\int_r^{+\infty} \phi(x)\mathrm{d}x\right] \mathrm{d}r - \int_{-\infty}^{0} \left[\int_{-\infty}^{r} \phi(x)\mathrm{d}x\right] \mathrm{d}r \\
&= \int_0^{+\infty} \left[\int_0^{x} \phi(x)\mathrm{d}r\right] \mathrm{d}x - \int_{-\infty}^{0} \left[\int_x^{0} \phi(x)\mathrm{d}r\right] \mathrm{d}x \\
&= \int_0^{+\infty} x\phi(x)\mathrm{d}x + \int_{-\infty}^{0} x\phi(x)\mathrm{d}x \\
&= \int_{-\infty}^{+\infty} x\phi(x)\mathrm{d}x.
\end{aligned}$$

The theorem is proved.

Example 4.7. Let ξ be a uniformly distributed random variable on the interval $[a, b]$. Then its expected value is

$$E[\xi] = \int_a^b \frac{x}{b-a}\mathrm{d}x = \frac{a+b}{2}$$

.

Example 4.8. Let ξ be an exponentially distributed random variable $\mathcal{EXP}(\beta)$. Then its expected value is

$$E[\xi] = \int_0^{+\infty} \frac{x}{\beta} \exp\left(-\frac{x}{\beta}\right) \mathrm{d}x = \beta.$$

Example 4.9. Let ξ be a normally distributed random variable $\mathcal{N}(\mu, \sigma^2)$. Then its expected value is

$$E[\xi] = \int_{-\infty}^{+\infty} \frac{x}{\sigma\sqrt{2\pi}} \exp\left(-\frac{(x-\mu)^2}{2\sigma^2}\right) \mathrm{d}x = \mu.$$

Critical Values

Let ξ be a random variable. In order to measure it, we may use its expected value. Alternately, we may employ α-optimistic value and α-pessimistic value as a ranking measure.

Definition 4.10. *Let ξ be a random variable, and $\alpha \in (0,1]$. Then*

$$\xi_{\sup}(\alpha) = \sup\left\{r \mid \Pr\{\xi \geq r\} \geq \alpha\right\} \tag{4.20}$$

is called the α-optimistic value of ξ; and

$$\xi_{\inf}(\alpha) = \inf\left\{r \mid \Pr\{\xi \leq r\} \geq \alpha\right\} \tag{4.21}$$

is called the α-pessimistic value of ξ.

This means that the random variable ξ will reach upwards of the α-optimistic value $\xi_{\sup}(\alpha)$ at least α of time, and will be below the α-pessimistic value $\xi_{\inf}(\alpha)$ at least α of time.

Theorem 4.7. *Let ξ be a random variable. Then we have*
(a) $\xi_{\inf}(\alpha)$ is an increasing and left-continuous function of α;
(b) $\xi_{\sup}(\alpha)$ is a decreasing and left-continuous function of α.

Proof: (a) It is easy to prove that $\xi_{\inf}(\alpha)$ is an increasing function of α. Next, we prove the left-continuity of $\xi_{\inf}(\alpha)$ with respect to α. Let $\{\alpha_i\}$ be an arbitrary sequence of positive numbers such that $\alpha_i \uparrow \alpha$. Then $\{\xi_{\inf}(\alpha_i)\}$ is an increasing sequence. If the limitation is equal to $\xi_{\inf}(\alpha)$, then the left-continuity is proved. Otherwise, there exists a number z^* such that

$$\lim_{i\to\infty} \xi_{\inf}(\alpha_i) < z^* < \xi_{\inf}(\alpha).$$

Thus $\Pr\{\xi \leq z^*\} \geq \alpha_i$ for each i. Letting $i \to \infty$, we get $\Pr\{\xi \leq z^*\} \geq \alpha$. Hence $z^* \geq \xi_{\inf}(\alpha)$. A contradiction proves the left-continuity of $\xi_{\inf}(\alpha)$ with respect to α. The part (b) may be proved similarly.

Ranking Criteria

Let ξ and η be two random variables. Different from the situation of real numbers, there does not exist a natural ordership in a random world. Thus an important problem appearing in this area is how to rank random variables. Here we give four ranking criteria.

Expected Value Criterion: We say $\xi > \eta$ if and only if $E[\xi] > E[\eta]$, where E is the expected value operator of random variables.

Optimistic Value Criterion: We say $\xi > \eta$ if and only if, for some predetermined confidence level $\alpha \in (0,1]$, we have $\xi_{\sup}(\alpha) > \eta_{\sup}(\alpha)$, where $\xi_{\sup}(\alpha)$ and $\eta_{\sup}(\alpha)$ are the α-optimistic values of ξ and η, respectively.

Pessimistic Value Criterion: We say $\xi > \eta$ if and only if, for some predetermined confidence level $\alpha \in (0,1]$, we have $\xi_{\inf}(\alpha) > \eta_{\inf}(\alpha)$, where $\xi_{\inf}(\alpha)$ and $\eta_{\inf}(\alpha)$ are the α-pessimistic values of ξ and η, respectively.

Probability Criterion: We say $\xi > \eta$ if and only if $\Pr\{\xi \geq \overline{r}\} > \Pr\{\eta \geq \overline{r}\}$ for some predetermined level $\overline{r}$.

Random Number Generation

Random number generation is a very important issue in Monte Carlo simulation. Generally, let ξ be a random variable with a probability distribution $\Phi(\cdot)$. Since $\Phi(\cdot)$ is an increasing function, the inverse function $\Phi^{-1}(\cdot)$ is defined on $[0,1]$. Assume that u is a uniformly distributed random variable on the interval $[0,1]$. Then we have

$$\Pr\left\{\Phi^{-1}(u) \leq y\right\} = \Pr\left\{u \leq \Phi(y)\right\} = \Phi(y) \tag{4.22}$$

which proves that the variable $\xi = \Phi^{-1}(u)$ has the probability distribution $\Phi(\cdot)$. In order to get a random variable ξ with probability distribution $\Phi(\cdot)$, we can produce a uniformly distributed random variable u from the interval $[0,1]$, and ξ is assigned to be $\Phi^{-1}(u)$. The above process is called the *inverse transform method.* But for the main known distributions, instead of using the inverse transform method, we have direct generating processes. For detailed expositions, the interested readers may consult Fishman [67], Law and Kelton [147], Bratley et al. [23], Rubinstein [268], and Liu [181]. Here we give some generating methods for probability distributions frequently used in this book.

The subfunction of generating pseudorandom numbers has been provided by the C library for any type of computer, defined as

```
int rand(void)
```

which produces a pseudorandom integer between 0 and RAND_MAX, where RAND_MAX is defined in stdlib.h as $2^{15}-1$. Thus the uniform distribution, exponential distribution, and normal distribution can be generated by the following way:

Algorithm 4.1 (Uniform Distribution $\mathcal{U}(a,b)$)
Step 1. $u = \text{rand}(\,)$.
Step 2. $u \leftarrow u/\text{RAND\_MAX}$.
Step 3. Return $a + u(b-a)$.

Algorithm 4.2 (Exponential Distribution $\mathcal{EXP}(\beta)$)
Step 1. Generate u from $\mathcal{U}(0,1)$.
Step 2. Return $-\beta\ln(u)$.

Algorithm 4.3 (Normal Distribution $\mathcal{N}(\mu,\sigma^2)$)
Step 1. Generate μ_1 and μ_2 from $\mathcal{U}(0,1)$.
Step 2. $y=[-2\ln(\mu_1)]^{\frac{1}{2}}\sin(2\pi\mu_2)$.
Step 3. Return $\mu+\sigma y$.

4.2 Expected Value Model

The first type of stochastic programming is the so-called *expected value model* (EVM), which optimizes some expected objective function subject to some expected constraints, for example, minimizing expected cost, maximizing expected profit, and so forth.

Now let us recall the well-known newsboy problem in which a boy operating a news stall has to determine the number x of newspapers to order in advance from the publisher at a cost of \$$c$/newspaper every day. It is known that the selling price is \$$a$/newspaper. However, if the newspapers are not sold at the end of the day, then the newspapers have a small value of \$$b$/newspaper at the recycling center. Assume that the demand for newspapers is denoted by ξ in a day, then the number of newspapers at the end of the day is clearly $x-\xi$ if $x>\xi$, or 0 if $x\le\xi$. Thus the profit of the newsboy should be

$$f(x,\xi)=\begin{cases}(a-c)x, & \text{if } x\le\xi\\ (b-c)x+(a-b)\xi, & \text{if } x>\xi.\end{cases}$$

In practice, the demand ξ for newspapers is usually a stochastic variable, so is the profit function $f(x,\xi)$. Since we cannot predict how profitable the decision of ordering x newspapers will actually be, a natural idea is to employ the expected profit $E[f(x,\xi)]$. The newsboy problem is related to determining the optimal integer number x of newspapers such that the expected profit $E[f(x,\xi)]$ achieves the maximal value, i.e.,

$$\begin{cases}\max E[f(x,\xi)]\\ \text{subject to:}\\ \qquad x\ge 0,\quad \text{integer}.\end{cases}$$

This is a typical example of EVM. Generally, if we want to find a decision with maximum expected return subject to some expected constraints, then we have the following EVM,

$$\begin{cases} \max E[f(\boldsymbol{x},\boldsymbol{\xi})] \\ \text{subject to:} \\ \qquad E[g_j(\boldsymbol{x},\boldsymbol{\xi})] \le 0,\ j = 1,2,\cdots,p \end{cases} \tag{4.23}$$

where $\boldsymbol{x}$ is a decision vector, $\boldsymbol{\xi}$ is a stochastic vector, $f(\boldsymbol{x},\boldsymbol{\xi})$ is the return function, $g_j(\boldsymbol{x},\boldsymbol{\xi})$ are stochastic constraint functions for $j = 1,2,\cdots,p$.

Definition 4.11. *A solution $\boldsymbol{x}$ is feasible if and only if $E[g_j(\boldsymbol{x},\boldsymbol{\xi})] \le 0$ for $j = 1,2,\cdots,p$. A feasible solution $\boldsymbol{x}^*$ is an optimal solution to EVM (4.23) if $E[f(\boldsymbol{x}^*,\boldsymbol{\xi})] \ge E[f(\boldsymbol{x},\boldsymbol{\xi})]$ for any feasible solution $\boldsymbol{x}$.*

In many cases, there are multiple objectives. Thus we have to employ the following expected value multiobjective programming (EVMOP),

$$\begin{cases} \max \left[E[f_1(\boldsymbol{x},\boldsymbol{\xi})], E[f_2(\boldsymbol{x},\boldsymbol{\xi})], \cdots, E[f_m(\boldsymbol{x},\boldsymbol{\xi})]\right] \\ \text{subject to:} \\ \qquad E[g_j(\boldsymbol{x},\boldsymbol{\xi})] \le 0,\ j = 1,2,\cdots,p \end{cases} \tag{4.24}$$

where $f_i(\boldsymbol{x},\boldsymbol{\xi})$ are return functions for $i = 1,2,\cdots,m$.

Definition 4.12. *A feasible solution $\boldsymbol{x}^*$ is said to be a Pareto solution to EVMOP (4.24) if there is no feasible solution $\boldsymbol{x}$ such that*

$$E[f_i(\boldsymbol{x},\boldsymbol{\xi})] \ge E[f_i(\boldsymbol{x}^*,\boldsymbol{\xi})], \quad i = 1,2,\cdots,m \tag{4.25}$$

and $E[f_j(\boldsymbol{x},\boldsymbol{\xi})] > E[f_j(\boldsymbol{x}^*,\boldsymbol{\xi})]$ for at least one index j.

We can also formulate a stochastic decision system as an expected value goal programming (EVGP) according to the priority structure and target levels set by the decision-maker:

$$\begin{cases} \min \sum\limits_{j=1}^{l} P_j \sum\limits_{i=1}^{m} (u_{ij} d_i^+ \vee 0 + v_{ij} d_i^- \vee 0) \\ \text{subject to:} \\ \qquad E[f_i(\boldsymbol{x},\boldsymbol{\xi})] - b_i = d_i^+,\ i = 1,2,\cdots,m \\ \qquad b_i - E[f_i(\boldsymbol{x},\boldsymbol{\xi})] = d_i^-,\ i = 1,2,\cdots,m \\ \qquad E[g_j(\boldsymbol{x},\boldsymbol{\xi})] \le 0, \qquad j = 1,2,\cdots,p \end{cases} \tag{4.26}$$

where P_j is the preemptive priority factor which expresses the relative importance of various goals, $P_j \gg P_{j+1}$, for all j, u_{ij} is the weighting factor corresponding to positive deviation for goal i with priority j assigned, v_{ij} is the weighting factor corresponding to negative deviation for goal i with priority j assigned, $d_i^+ \vee 0$ is the positive deviation from the target of goal i, $d_i^- \vee 0$ is the negative deviation from the target of goal i, f_i is a function in goal constraints, g_j is a function in real constraints, b_i is the target value according to goal i, l is the number of priorities, m is the number of goal constraints, and p is the number of real constraints.

4.3 Chance-Constrained Programming

As the second type of stochastic programming developed by Charnes and Cooper [37], chance-constrained programming (CCP) offers a powerful means of modeling stochastic decision systems with assumption that the stochastic constraints will hold at least α of time, where α is referred to as the *confidence level* provided as an appropriate safety margin by the decision-maker. After that, Liu [174] generalized CCP to the case with not only stochastic constraints but also stochastic objectives.

Assume that $\boldsymbol{x}$ is a decision vector, $\boldsymbol{\xi}$ is a stochastic vector, $f(\boldsymbol{x}, \boldsymbol{\xi})$ is a return function, and $g_j(\boldsymbol{x}, \boldsymbol{\xi})$ are stochastic constraint functions, $j = 1, 2, \cdots, p$. Since the stochastic constraints $g_j(\boldsymbol{x}, \boldsymbol{\xi}) \leq 0, j = 1, 2, \cdots, p$ do not define a deterministic feasible set, it is desired that the stochastic constraints hold with a confidence level α. Thus we have a chance constraint as follows,

$$\Pr\left\{g_j(\boldsymbol{x}, \boldsymbol{\xi}) \leq 0, j = 1, 2, \cdots, p\right\} \geq \alpha \tag{4.27}$$

which is called a joint chance constraint.

Definition 4.13. *A point $\boldsymbol{x}$ is called feasible if and only if the probability measure of the event $\{g_j(\boldsymbol{x}, \boldsymbol{\xi}) \leq 0, j = 1, 2, \cdots, p\}$ is at least α.*

In other words, the constraints will be violated at most $(1 - \alpha)$ of time. Sometimes, the joint chance constraint is separately considered as

$$\Pr\{g_j(\boldsymbol{x}, \boldsymbol{\xi}) \leq 0\} \geq \alpha_j, \quad j = 1, 2, \cdots, p \tag{4.28}$$

which is referred to as a separate chance constraint.

Maximax Chance-Constrained Programming

In a stochastic environment, in order to maximize the optimistic return with a given confidence level subject to some chance constraint, Liu [174] gave the following CCP:

$$\begin{cases} \max\limits_{\boldsymbol{x}} \max\limits_{\overline{f}} \overline{f} \\ \text{subject to:} \\ \qquad \Pr\left\{f(\boldsymbol{x}, \boldsymbol{\xi}) \geq \overline{f}\right\} \geq \beta \\ \qquad \Pr\left\{g_j(\boldsymbol{x}, \boldsymbol{\xi}) \leq 0, j = 1, 2, \cdots, p\right\} \geq \alpha \end{cases} \tag{4.29}$$

where α and β are the predetermined confidence levels, and $\max \overline{f}$ is the β-optimistic return.

In practice, we may have multiple objectives. Thus we have to employ the following chance-constrained multiobjective programming (CCMOP),

$$\begin{cases} \max\limits_{\boldsymbol{x}} \left[\max\limits_{\overline{f}_1} \overline{f}_1, \max\limits_{\overline{f}_2} \overline{f}_2, \cdots, \max\limits_{\overline{f}_m} \overline{f}_m \right] \\ \text{subject to:} \\ \qquad \Pr\left\{ f_i(\boldsymbol{x}, \boldsymbol{\xi}) \ge \overline{f}_i \right\} \ge \beta_i, \ i = 1, 2, \cdots, m \\ \qquad \Pr\left\{ g_j(\boldsymbol{x}, \boldsymbol{\xi}) \le 0 \right\} \ge \alpha_j, \ \ j = 1, 2, \cdots, p \end{cases} \tag{4.30}$$

where $\alpha_1, \alpha_2, \cdots, \alpha_p, \beta_1, \beta_2, \cdots, \beta_m$ are the predetermined confidence levels, and $\max \overline{f}_i$ are the β_i-optimistic values to the ith return functions $f_i(\boldsymbol{x}, \boldsymbol{\xi})$, $i = 1, 2, \cdots, m$, respectively.

Sometimes, we may formulate a stochastic decision system as a chance-constrained goal programming (CCGP) according to the priority structure and target levels set by the decision-maker:

$$\begin{cases} \min\limits_{\boldsymbol{x}} \sum\limits_{j=1}^{l} P_j \sum\limits_{i=1}^{m} \left(u_{ij} \left(\min\limits_{d_i^+} d_i^+ \vee 0 \right) + v_{ij} \left(\min\limits_{d_i^-} d_i^- \vee 0 \right) \right) \\ \text{subject to:} \\ \qquad \Pr\left\{ f_i(\boldsymbol{x}, \boldsymbol{\xi}) - b_i \le d_i^+ \right\} \ge \beta_i^+, \ i = 1, 2, \cdots, m \\ \qquad \Pr\left\{ b_i - f_i(\boldsymbol{x}, \boldsymbol{\xi}) \le d_i^- \right\} \ge \beta_i^-, \ i = 1, 2, \cdots, m \\ \qquad \Pr\left\{ g_j(\boldsymbol{x}, \boldsymbol{\xi}) \le 0 \right\} \ge \alpha_j, \qquad j = 1, 2, \cdots, p \end{cases} \tag{4.31}$$

where P_j is the preemptive priority factor which expresses the relative importance of various goals, $P_j \gg P_{j+1}$, for all j, u_{ij} is the weighting factor corresponding to positive deviation for goal i with priority j assigned, v_{ij} is the weighting factor corresponding to negative deviation for goal i with priority j assigned, $\min d_i^+ \vee 0$ is the β_i^+-optimistic positive deviation from the target of goal i, $\min d_i^- \vee 0$ is the β_i^--optimistic negative deviation from the target of goal i, f_i is a function in goal constraints, g_j is a function in system constraints, b_i is the target value according to goal i, l is the number of priorities, m is the number of goal constraints, and p is the number of system constraints.

Remark 4.1. In a deterministic goal programming, at most one of positive deviation and negative deviation takes a positive value. However, for a CCGP, it is possible that both of them are positive.

Minimax Chance-Constrained Programming

In a stochastic environment, in order to maximize the pessimistic return with a given confidence level subject to some chance constraint, Liu [181] provided the following minimax CCP model:

$$\begin{cases} \max\limits_{\boldsymbol{x}} \min\limits_{\overline{f}} \overline{f} \\ \text{subject to:} \\ \qquad \Pr\left\{f(\boldsymbol{x},\boldsymbol{\xi}) \le \overline{f}\right\} \ge \beta \\ \qquad \Pr\left\{g_j(\boldsymbol{x},\boldsymbol{\xi}) \le 0, j=1,2,\cdots,p\right\} \ge \alpha \end{cases} \tag{4.32}$$

where α and β are the given confidence levels, and $\min \overline{f}$ is the β-pessimistic return.

If there are multiple objectives, then we may employ the following minimax CCMOP,

$$\begin{cases} \max\limits_{\boldsymbol{x}} \left[\min\limits_{\overline{f}_1} \overline{f}_1,\ \min\limits_{\overline{f}_2} \overline{f}_2,\ \cdots,\ \min\limits_{\overline{f}_m} \overline{f}_m\right] \\ \text{subject to:} \\ \qquad \Pr\left\{f_i(\boldsymbol{x},\boldsymbol{\xi}) \le \overline{f}_i\right\} \ge \beta_i,\ i=1,2,\cdots,m \\ \qquad \Pr\left\{g_j(\boldsymbol{x},\boldsymbol{\xi}) \le 0\right\} \ge \alpha_j,\ \ j=1,2,\cdots,p \end{cases} \tag{4.33}$$

where α_j and β_i are confidence levels, and $\min \overline{f}_i$ are the β_i-pessimistic values to the return functions $f_i(\boldsymbol{x},\boldsymbol{\xi})$, $i=1,2,\cdots,m$, $j=1,2,\cdots,p$, respectively.

We can also formulate a stochastic decision system as a minimax CCGP according to the priority structure and target levels set by the decision-maker:

$$\begin{cases} \min\limits_{\boldsymbol{x}} \sum\limits_{j=1}^{l} P_j \sum\limits_{i=1}^{m} \left[u_{ij}\left(\max\limits_{d_i^+} d_i^+ \vee 0\right) + v_{ij}\left(\max\limits_{d_i^-} d_i^- \vee 0\right)\right] \\ \text{subject to:} \\ \qquad \Pr\left\{f_i(\boldsymbol{x},\boldsymbol{\xi}) - b_i \ge d_i^+\right\} \ge \beta_i^+,\ i=1,2,\cdots,m \\ \qquad \Pr\left\{b_i - f_i(\boldsymbol{x},\boldsymbol{\xi}) \ge d_i^-\right\} \ge \beta_i^-,\ i=1,2,\cdots,m \\ \qquad \Pr\left\{g_j(\boldsymbol{x},\boldsymbol{\xi}) \le 0\right\} \ge \alpha_j,\qquad\quad j=1,2,\cdots,p \end{cases} \tag{4.34}$$

where P_j is the preemptive priority factor which expresses the relative importance of various goals, $P_j \gg P_{j+1}$, for all j, u_{ij} is the weighting factor corresponding to positive deviation for goal i with priority j assigned, v_{ij} is the weighting factor corresponding to negative deviation for goal i with priority j assigned, $\max d_i^+ \vee 0$ is the β_i^+-pessimistic positive deviation from the target of goal i, $\max d_i^- \vee 0$ is the β_i^--pessimistic negative deviation from the target of goal i, f_i is a function in goal constraints, g_j is a function in system constraints, b_i is the target value according to goal i, l is the number of priorities, m is the number of goal constraints, and p is the number of system constraints.

Deterministic Equivalents

The traditional solution methods require conversion of the chance constraints to their respective deterministic equivalents. As we know, this process is usually hard to perform and only successful for some special cases. Let us consider the following form of chance constraint,

$$\Pr\{g(\boldsymbol{x},\boldsymbol{\xi}) \le 0\} \ge \alpha. \tag{4.35}$$

It is clear that

(a)the chance constraints (4.28) are a set of form (4.35);
(b)the stochastic objective constraint $\Pr\{f(\boldsymbol{x},\boldsymbol{\xi}) \ge \overline{f}\} \ge \beta$ coincides with the form (4.35) by defining $g(\boldsymbol{x},\boldsymbol{\xi}) = \overline{f} - f(\boldsymbol{x},\boldsymbol{\xi})$;
(c)the stochastic objective constraint $\Pr\{f(\boldsymbol{x},\boldsymbol{\xi}) \le \overline{f}\} \ge \beta$ coincides with the form (4.35) by defining $g(\boldsymbol{x},\boldsymbol{\xi}) = f(\boldsymbol{x},\boldsymbol{\xi}) - \overline{f}$;
(d)the stochastic goal constraints $\Pr\{b - f(\boldsymbol{x},\boldsymbol{\xi}) \le d^-\} \ge \beta$ and $\Pr\{f(\boldsymbol{x},\boldsymbol{\xi}) - b \le d^+\} \ge \beta$ coincide with the form (4.35) by defining $g(\boldsymbol{x},\boldsymbol{\xi}) = b - f(\boldsymbol{x},\boldsymbol{\xi}) - d^-$ and $g(\boldsymbol{x},\boldsymbol{\xi}) = f(\boldsymbol{x},\boldsymbol{\xi}) - b - d^+$, respectively; and
(e)the stochastic goal constraints $\Pr\{b - f(\boldsymbol{x},\boldsymbol{\xi}) \ge d^-\} \ge \beta$ and $\Pr\{f(\boldsymbol{x},\boldsymbol{\xi}) - b \ge d^+\} \ge \beta$ coincide with the form (4.35) by defining $g(\boldsymbol{x},\boldsymbol{\xi}) = f(\boldsymbol{x},\boldsymbol{\xi}) + d^- - b$ and $g(\boldsymbol{x},\boldsymbol{\xi}) = b - f(\boldsymbol{x},\boldsymbol{\xi}) + d^+$, respectively.

This section summarizes some known results.

Theorem 4.8. *Assume that the stochastic vector $\boldsymbol{\xi}$ degenerates to a random variable ξ with probability distribution Φ, and the function $g(\boldsymbol{x},\boldsymbol{\xi})$ has the form $g(\boldsymbol{x},\boldsymbol{\xi}) = h(\boldsymbol{x}) - \xi$. Then $\Pr\{g(\boldsymbol{x},\boldsymbol{\xi}) \le 0\} \ge \alpha$ if and only if $h(\boldsymbol{x}) \le K_\alpha$, where K_α is the maximal number such that $\Pr\{K_\alpha \le \xi\} \ge \alpha$.*

Proof: The assumption implies that $\Pr\{g(\boldsymbol{x},\boldsymbol{\xi}) \le 0\} \ge \alpha$ can be written in the following form,

$$\Pr\{h(\boldsymbol{x}) \le \xi\} \ge \alpha. \tag{4.36}$$

For each given confidence level α ($0 < \alpha \le 1$), let K_α be the maximal number (maybe $+\infty$) such that

$$\Pr\{K_\alpha \le \xi\} \ge \alpha. \tag{4.37}$$

Note that the probability $\Pr\{K_\alpha \le \xi\}$ will increase if K_α is replaced with a smaller number. Hence $\Pr\{h(\boldsymbol{x}) \le \xi\} \ge \alpha$ if and only if $h(\boldsymbol{x}) \le K_\alpha$.

Remark 4.2. For a continuous random variable ξ, the equation $\Pr\{K_\alpha \le \xi\} = 1 - \Phi(K_\alpha)$ always holds, and we have, by (4.37),

$$K_\alpha = \Phi^{-1}(1-\alpha) \tag{4.38}$$

where Φ^{-1} is the inverse function of Φ.

Example 4.10. Assume that we have the following chance constraint,

$$\begin{cases} \Pr\{3x_1 + 4x_2 \le \xi_1\} \ge 0.8 \\ \Pr\{x_1^2 - x_2^3 \le \xi_2\} \ge 0.9 \end{cases} \tag{4.39}$$

where ξ_1 is an exponentially distributed random variable $\mathcal{EXP}(2)$ whose probability distribution is denoted by Φ_1, and ξ_2 is a normally distributed random variable $\mathcal{N}(2,1)$ whose probability distribution is denoted by Φ_2. It follows from Theorem 4.8 that the chance constraint (4.39) is equivalent to

$$\begin{cases} 3x_1 + 4x_2 \le \Phi_1^{-1}(1-0.8) = 0.446 \\ x_1^2 - x_2^3 \le \Phi_2^{-1}(1-0.9) = 0.719. \end{cases}$$

Theorem 4.9. *Assume that the stochastic vector* $\boldsymbol{\xi} = (a_1, a_2, \cdots, a_n, b)$ *and the function* $g(\boldsymbol{x}, \boldsymbol{\xi})$ *has the form* $g(\boldsymbol{x}, \boldsymbol{\xi}) = a_1x_1 + a_2x_2 + \cdots + a_nx_n - b$. *If* a_i *and* b *are assumed to be independently normally distributed random variables, then* $\Pr\{g(\boldsymbol{x}, \boldsymbol{\xi}) \le 0\} \ge \alpha$ *if and only if*

$$\sum_{i=1}^{n} E[a_i]x_i + \Phi^{-1}(\alpha)\sqrt{\sum_{i=1}^{n} V[a_i]x_i^2 + V[b]} \le E[b] \tag{4.40}$$

where Φ *is the standardized normal distribution function.*

Proof: The chance constraint $\Pr\{g(\boldsymbol{x}, \boldsymbol{\xi}) \le 0\} \ge \alpha$ can be written in the following form,

$$\Pr\left\{\sum_{i=1}^{n} a_ix_i \le b\right\} \ge \alpha. \tag{4.41}$$

Since a_i and b are assumed to be independently normally distributed random variables, the quantity

$$y = \sum_{i=1}^{n} a_ix_i - b$$

is also normally distributed with the following expected value and variance,

$$E[y] = \sum_{i=1}^{n} E[a_i]x_i - E[b],$$
$$V[y] = \sum_{i=1}^{n} V[a_i]x_i^2 + V[b].$$

We note that

$$\frac{\sum_{i=1}^{n} a_ix_i - b - \left(\sum_{i=1}^{n} E[a_i]x_i - E[b]\right)}{\sqrt{\sum_{i=1}^{n} V[a_i]x_i^2 + V[b]}}$$

must be standardized normally distributed. Since the inequality $\sum_{i=1}^n a_i x_i \le b$ is equivalent to

$$\frac{\sum\limits_{i=1}^n a_i x_i - b - \left(\sum\limits_{i=1}^n E[a_i]x_i - E[b]\right)}{\sqrt{\sum\limits_{i=1}^n V[a_i]x_i^2 + V[b]}} \le -\frac{\sum\limits_{i=1}^n E[a_i]x_i - E[b]}{\sqrt{\sum\limits_{i=1}^n V[a_i]x_i^2 + V[b]}},$$

the chance constraint (4.41) is equivalent to

$$\Pr\left\{\eta \le -\frac{\sum\limits_{i=1}^n E[a_i]x_i - E[b]}{\sqrt{\sum\limits_{i=1}^n V[a_i]x_i^2 + V[b]}}\right\} \ge \alpha \tag{4.42}$$

where η is the standardized normally distributed random variable. Then the chance constraint (4.42) holds if and only if

$$\Phi^{-1}(\alpha) \le -\frac{\sum\limits_{i=1}^n E[a_i]x_i - E[b]}{\sqrt{\sum\limits_{i=1}^n V[a_i]x_i^2 + V[b]}}. \tag{4.43}$$

That is, the deterministic equivalent of chance constraint is (4.40). The theorem is proved.

Example 4.11. Suppose that the chance constraint set has the following form,

$$\Pr\{a_1 x_1 + a_2 x_2 + a_2 x_3 \le b\} \ge 0.95 \tag{4.44}$$

where a_1, a_2, a_3, and b are normally distributed random variables $\mathcal{N}(1,1)$, $\mathcal{N}(2,1)$, $\mathcal{N}(3,1)$, and $\mathcal{N}(4,1)$, respectively. Then the formula (4.40) yields the deterministic equivalent of (4.44) as follows,

$$x_1 + 2x_2 + 3x_3 + 1.645\sqrt{x_1^2 + x_2^2 + x_3^2 + 1} \le 4$$

by the fact that $\Phi^{-1}(0.95) = 1.645$.

4.4 Dependent-Chance Programming

In practice, there usually exist multiple events in a complex stochastic decision system. Sometimes, the decision-maker wishes to maximize the probabilities of meeting these events. In order to model this type of stochastic decision system, Liu [166] provided the third type of stochastic programming, called *dependent-chance programming* (DCP), in which the underlying philosophy is based on selecting the decision with maximal chance to meet the event.

DCP theory breaks the concept of feasible set and replaces it with uncertain environment. Roughly speaking, DCP involves maximizing chance functions of events in an uncertain environment. In deterministic model, EVM and CCP, the feasible set is essentially assumed to be deterministic after the real problem is modeled. That is, an optimal solution is given regardless of whether it can be performed in practice. However, the given solution may be impossible to perform if the realization of uncertain parameter is unfavorable. Thus DCP theory never assumes that the feasible set is deterministic. In fact, DCP is constructed in an uncertain environment. This special feature of DCP is very different from the other existing types of stochastic programming. However, such problems do exist in the real world.

Now we introduce the concepts of uncertain environment, event and chance function, and discuss the principle of uncertainty, thus offering a spectrum of DCP models. We will take a supply system, represented by Figure 4.1 as the background.

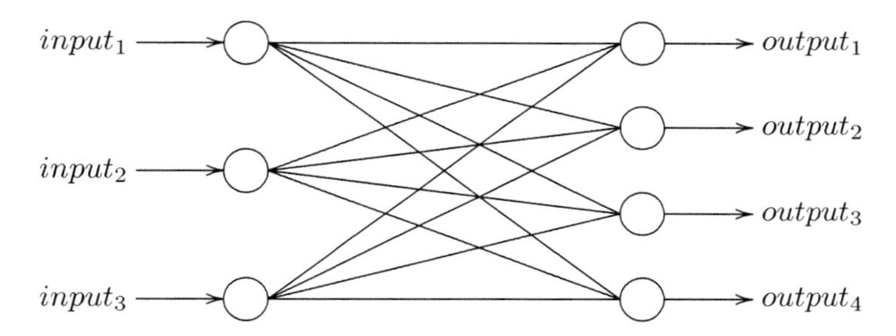

Fig. 4.1 A Supply System

As an illustrative example, in Figure 4.1 there are 3 inputs representing 3 locations of resources and 4 outputs representing the demands of 4 users. We must answer the following supply problem: What is the appropriate combination of resources such that certain goals of supply are achieved?

In order to obtain the appropriate combination of resources for the supply problem, we use 12 decision variables $x_1, x_2, \cdots, x_{12}$ to represent an action, where x_1, x_2, x_3, x_4 are quantities ordered from $input_1$ to outputs 1,2,3,4 respectively; x_5, x_6, x_7, x_8 from $input_2$; $x_9, x_{10}, x_{11}, x_{12}$ from $input_3$. In practice, some variables may vanish due to some physical constraints.

We note that the inputs are available outside resources. Thus they have their own properties. For example, the capacities of resources are finite. Let ξ_1, ξ_2, ξ_3 be the maximum quantities supplied by the three resources. Then we have the following constraint,

$$\begin{cases} x_1^+ + x_2^+ + x_3^+ + x_4^+ \le \xi_1 \\ x_5^+ + x_6^+ + x_7^+ + x_8^+ \le \xi_2 \\ x_9^+ + x_{10}^+ + x_{11}^+ + x_{12}^+ \le \xi_3 \\ x_i \ge 0, \quad i = 1, 2, \cdots, 12 \end{cases} \tag{4.45}$$

which represents that the quantities ordered from the resources are nonnegative and cannot exceed the maximum quantities, where x_i^+ represents x_i if x_i takes positive value, and vanishes otherwise. This means that the decision variable $x_i = 0$ must be able to perform for any realization of stochastic resources.

If at least one of ξ_1, ξ_2, and ξ_3 is really stochastic, then the constraint (4.45) is uncertain because we cannot make a decision such that it can be performed certainly before knowing the realization of ξ_1, ξ_2, and ξ_3. We will call this type of constraint the uncertain environment, and in this case the stochastic environment.

Definition 4.14. *By uncertain environment we mean the following stochastic constraint,*

$$g_j(\boldsymbol{x}, \boldsymbol{\xi}) \le 0, \quad j = 1, 2, \cdots, p, \tag{4.46}$$

where $\boldsymbol{x}$ is a decision vector, and $\boldsymbol{\xi}$ is a stochastic vector.

In the supply system, we should satisfy the demands of the 4 users, marked by c_1, c_2, c_3, and c_4. Then we have the following four events:

$$x_1 + x_5 + x_9 = c_1, \quad x_2 + x_6 + x_{10} = c_2,$$
$$x_3 + x_7 + x_{11} = c_3, \; x_4 + x_8 + x_{12} = c_4.$$

These equalities mean that the decision should satisfy the demands of users. Generally, an event is defined as follows.

Definition 4.15. *By event we mean a system of stochastic inequalities,*

$$h_k(\boldsymbol{x}, \boldsymbol{\xi}) \le 0, \quad k = 1, 2, \cdots, q \tag{4.47}$$

where $\boldsymbol{x}$ is a decision vector, and $\boldsymbol{\xi}$ is a stochastic vector.

In view of the uncertainty of this system, we are not sure whether a decision can be performed before knowing the realization of stochastic variables. Thus we wish to employ the following chance functions to evaluate these four events,

$$f_1(\boldsymbol{x}) = \Pr\{x_1 + x_5 + x_9 = c_1\}, \;\; f_2(\boldsymbol{x}) = \Pr\{x_2 + x_6 + x_{10} = c_2\},$$
$$f_3(\boldsymbol{x}) = \Pr\{x_3 + x_7 + x_{11} = c_3\}, \; f_4(\boldsymbol{x}) = \Pr\{x_4 + x_8 + x_{12} = c_4\},$$

subject to the uncertain environment (4.45).

Definition 4.16. *The chance function of an event $\mathcal{E}$ characterized by (4.47) is defined as the probability measure of the event, i.e.,*

$$f(\boldsymbol{x}) = \Pr\{h_k(\boldsymbol{x}, \boldsymbol{\xi}) \le 0, k = 1, 2, \cdots, q\} \tag{4.48}$$

subject to the uncertain environment (4.46).

Usually, we hope to maximize the four chance functions $f_1(\boldsymbol{x})$, $f_2(\boldsymbol{x})$, $f_3(\boldsymbol{x})$ and $f_4(\boldsymbol{x})$. Here we remind the reader once more that the events like $x_1 + x_5 + x_9 = c_1$ do possess uncertainty because they are in an uncertain environment. *Any event is uncertain if it is in an uncertain environment!* This is an important law in the uncertain world. In fact, the randomness of the event is caused by the stochastic parameters ξ_1, ξ_2, ξ_3, and ξ_4 in the uncertain environment.

Until now we have formulated a stochastic programming model for the supply problem in an uncertain environment as follows,

$$\begin{cases} \max f_1(\boldsymbol{x}) = \Pr\{x_1 + x_5 + x_9 = c_1\} \\ \max f_2(\boldsymbol{x}) = \Pr\{x_2 + x_6 + x_{10} = c_2\} \\ \max f_3(\boldsymbol{x}) = \Pr\{x_3 + x_7 + x_{11} = c_3\} \\ \max f_4(\boldsymbol{x}) = \Pr\{x_4 + x_8 + x_{12} = c_4\} \\ \text{subject to:} \\ \qquad x_1^+ + x_2^+ + x_3^+ + x_4^+ \le \xi_1 \\ \qquad x_5^+ + x_6^+ + x_7^+ + x_8^+ \le \xi_2 \\ \qquad x_9^+ + x_{10}^+ + x_{11}^+ + x_{12}^+ \le \xi_3 \\ \qquad x_i \ge 0, \quad i = 1, 2, \cdots, 12 \end{cases} \tag{4.49}$$

where ξ_1, ξ_2, and ξ_3 are stochastic variables. In this stochastic programming model, some variables (for example, x_1, x_2, x_3, x_4) are stochastically dependent because they share a common uncertain resource ξ_1. This also implies that the chance functions are stochastically dependent. We will call the stochastic programming (4.49) *dependent-chance programming* (DCP).

Principle of Uncertainty

How do we compute the chance function of an event $\mathcal{E}$ in an uncertain environment? In order to answer this question, we first give some definitions.

Definition 4.17. *Let $r(x_1, x_2, \cdots, x_n)$ be an n-dimensional function. The ith decision variable x_i is said to be degenerate if*

$$r(x_1, \cdots, x_{i-1}, x_i', x_{i+1}, \cdots, x_n) = r(x_1, \cdots, x_{i-1}, x_i'', x_{i+1}, \cdots, x_n)$$

for any x_i' and x_i''; otherwise it is nondegenerate.

For example, $r(x_1, x_2, x_3, x_4, x_5) = (x_1 + x_3)/x_4$ is a 5-dimensional function. The variables x_1, x_3, x_4 are nondegenerate, but x_2 and x_5 are degenerate.

Definition 4.18. *Let $\mathcal{E}$ be an event $h_k(\boldsymbol{x},\boldsymbol{\xi}) \le 0$, $k = 1,2,\cdots,q$. The support of the event $\mathcal{E}$, denoted by $\mathcal{E}^*$, is defined as the set consisting of all nondegenerate decision variables of functions $h_k(\boldsymbol{x},\boldsymbol{\xi})$, $k = 1,2,\cdots,q$.*

For example, let $\boldsymbol{x} = (x_1, x_2, \cdots, x_{12})$ be a decision vector, and let $\mathcal{E}$ be an event characterized by $x_1 + x_5 + x_9 = c_1$ and $x_2 + x_6 + x_{10} = c_2$. It is clear that x_1, x_5, x_9 are nondegenerate variables of the function $x_1 + x_5 + x_9$, and x_2, x_6, x_{10} are nondegenerate variables of the function $x_2 + x_6 + x_{10}$. Thus the support $\mathcal{E}^*$ of the event $\mathcal{E}$ is $\{x_1, x_2, x_5, x_6, x_9, x_{10}\}$.

Definition 4.19. *The jth constraint $g_j(\boldsymbol{x},\boldsymbol{\xi}) \le 0$ is called an active constraint of the event $\mathcal{E}$ if the set of nondegenerate decision variables of $g_j(\boldsymbol{x},\boldsymbol{\xi})$ and the support $\mathcal{E}^*$ have nonempty intersection; otherwise it is inactive.*

Definition 4.20. *Let $\mathcal{E}$ be an event $h_k(\boldsymbol{x},\boldsymbol{\xi}) \le 0$, $k = 1,2,\cdots,q$ in the uncertain environment $g_j(\boldsymbol{x},\boldsymbol{\xi}) \le 0, j = 1,2,\cdots,p$. The dependent support of the event $\mathcal{E}$, denoted by $\mathcal{E}^{**}$, is defined as the set consisting of all nondegenerate decision variables of $h_k(\boldsymbol{x},\boldsymbol{\xi}), k = 1,2,\cdots,q$ and $g_j(\boldsymbol{x},\boldsymbol{\xi})$ in the active constraints to the event $\mathcal{E}$.*

Remark 4.3. It is obvious that $\mathcal{E}^* \subset \mathcal{E}^{**}$ holds.

Definition 4.21. *The jth constraint $g_j(\boldsymbol{x},\boldsymbol{\xi}) \le 0$ is called a dependent constraint of the event $\mathcal{E}$ if the set of nondegenerate decision variables of $g_j(\boldsymbol{x},\boldsymbol{\xi})$ and the dependent support $\mathcal{E}^{**}$ have nonempty intersection; otherwise it is independent.*

Remark 4.4. An active constraint must be a dependent constraint.

Definition 4.22. *Let $\mathcal{E}$ be an event $h_k(\boldsymbol{x},\boldsymbol{\xi}) \le 0$, $k = 1,2,\cdots,q$ in the uncertain environment $g_j(\boldsymbol{x},\boldsymbol{\xi}) \le 0, j = 1,2,\cdots,p$. For each decision $\boldsymbol{x}$ and realization $\boldsymbol{\xi}$, the event $\mathcal{E}$ is said to be consistent in the uncertain environment if the following two conditions hold: (i) $h_k(\boldsymbol{x},\boldsymbol{\xi}) \le 0$, $k = 1,2,\cdots,q$; and (ii) $g_j(\boldsymbol{x},\boldsymbol{\xi}) \le 0$, $j \in J$, where J is the index set of all dependent constraints.*

Intuitively, an event can be met by a decision provided that the decision meets both the event itself and the dependent constraints. We conclude it with the following principle of uncertainty.

Principle of Uncertainty: *The chance of a random event is the probability that the event is consistent in the uncertain environment.*

Assume that there are m events $\mathcal{E}_i$ characterized by $h_{ik}(\boldsymbol{x},\boldsymbol{\xi}) \le 0, k = 1,2,\cdots,q_i$ for $i = 1,2,\cdots,m$ in the uncertain environment $g_j(\boldsymbol{x},\boldsymbol{\xi}) \le 0, j = 1,2,\cdots,p$. The principle of uncertainty implies that the chance function of the ith event $\mathcal{E}_i$ in the uncertain environment is

$$f_i(\boldsymbol{x}) = \Pr\left\{\begin{array}{l} h_{ik}(\boldsymbol{x},\boldsymbol{\xi}) \le 0, k = 1,2,\cdots,q_i \\ g_j(\boldsymbol{x},\boldsymbol{\xi}) \le 0, j \in J_i \end{array}\right\} \tag{4.50}$$

where J_i are defined by

$$J_i = \{j \in \{1, 2, \cdots, p\} \mid g_j(\boldsymbol{x}, \boldsymbol{\xi}) \le 0 \text{ is a dependent constraint of } \mathcal{E}_i\}$$

for $i = 1, 2, \cdots, m$.

Remark 4.5. The principle of uncertainty is the basis of solution procedure of DCP that we shall encounter throughout the remainder of the book. However, the principle of uncertainty does not apply in all cases. For example, consider an event $x_1 \ge 6$ in the uncertain environment $x_1 - x_2 \le \xi_1, x_2 - x_3 \le \xi_2, x_3 \le \xi_3$. It follows from the principle of uncertainty that the chance of the event is $\Pr\{x_1 \ge 6, x_1 - x_2 \le \xi_1, x_2 - x_3 \le \xi_2\}$, which is clearly wrong because the realization of $x_3 \le \xi_3$ must be considered. Fortunately, such a case does not exist in real-life problems.

General Models

In this subsection, we consider the single-objective DCP. A typical DCP is represented as maximizing the chance function of an event subject to an uncertain environment,

$$\begin{cases} \max \Pr\{h_k(\boldsymbol{x}, \boldsymbol{\xi}) \le 0, k = 1, 2, \cdots, q\} \\ \text{subject to:} \\ \qquad g_j(\boldsymbol{x}, \boldsymbol{\xi}) \le 0, \quad j = 1, 2, \cdots, p \end{cases} \tag{4.51}$$

where $\boldsymbol{x}$ is an n-dimensional decision vector, $\boldsymbol{\xi}$ is a random vector of parameters, the system $h_k(\boldsymbol{x}, \boldsymbol{\xi}) \le 0$, $k = 1, 2, \cdots, q$ represents an event $\mathcal{E}$, and the constraints $g_j(\boldsymbol{x}, \boldsymbol{\xi}) \le 0$, $j = 1, 2, \cdots, p$ are an uncertain environment.

DCP (4.51) reads as "maximizing the probability of the random event $h_k(\boldsymbol{x}, \boldsymbol{\xi}) \le 0, k = 1, 2, \cdots, q$ subject to the uncertain environment $g_j(\boldsymbol{x}, \boldsymbol{\xi}) \le 0, j = 1, 2, \cdots, p$".

We now go back to the supply system. Assume that there is only one event $\mathcal{E}$ that satisfies the demand c_1 of *output*$_1$ (i.e., $x_1 + x_5 + x_9 = c_1$). If we want to find a decision $\boldsymbol{x}$ with maximum probability to meet the event $\mathcal{E}$, then we have the following DCP model,

$$\begin{cases} \max \Pr\{x_1 + x_5 + x_9 = c_1\} \\ \text{subject to:} \\ \qquad x_1^+ + x_2^+ + x_3^+ + x_4^+ \le \xi_1 \\ \qquad x_5^+ + x_6^+ + x_7^+ + x_8^+ \le \xi_2 \\ \qquad x_9^+ + x_{10}^+ + x_{11}^+ + x_{12}^+ \le \xi_3 \\ \qquad x_i \ge 0, i = 1, 2, \cdots, 12. \end{cases} \tag{4.52}$$

It is clear that the support of the event $\mathcal{E}$ is $\mathcal{E}^* = \{x_1, x_5, x_9\}$. If $x_1 \neq 0, x_5 \neq 0, x_9 \neq 0$, then the uncertain environment is

$$\begin{cases} x_1 + x_2 + x_3 + x_4 \le \xi_1 \\ x_5 + x_6 + x_7 + x_8 \le \xi_2 \\ x_9 + x_{10} + x_{11} + x_{12} \le \xi_3 \\ x_i \ge 0, i = 1, 2, \cdots, 12. \end{cases}$$

Thus the dependent support $\mathcal{E}^{**} = \{x_1, x_2, \cdots, x_{12}\}$, and all constraints are dependent constraints. It follows from the principle of uncertainty that the chance function of the event $\mathcal{E}$ is

$$f(\boldsymbol{x}) = \Pr\left\{\begin{array}{l} x_1 + x_5 + x_9 = c_1 \\ x_1 + x_2 + x_3 + x_4 \le \xi_1 \\ x_5 + x_6 + x_7 + x_8 \le \xi_2 \\ x_9 + x_{10} + x_{11} + x_{12} \le \xi_3 \\ x_i \ge 0, i = 1, 2, \cdots, 12 \end{array}\right\}.$$

If $x_1 = 0, x_5 \neq 0, x_9 \neq 0$, then the uncertain environment is

$$\begin{cases} 0 + x_2 + x_3 + x_4 \le \xi_1 \\ x_5 + x_6 + x_7 + x_8 \le \xi_2 \\ x_9 + x_{10} + x_{11} + x_{12} \le \xi_3 \\ x_i \ge 0, i = 1, 2, \cdots, 12. \end{cases}$$

Thus the dependent support $\mathcal{E}^{**} = \{x_5, x_6, \cdots, x_{12}\}$. It follows from the principle of uncertainty that the chance function of the event $\mathcal{E}$ is

$$f(\boldsymbol{x}) = \Pr\left\{\begin{array}{l} x_1 + x_5 + x_9 = c_1 \\ x_5 + x_6 + x_7 + x_8 \le \xi_2 \\ x_9 + x_{10} + x_{11} + x_{12} \le \xi_3 \\ x_i \ge 0, i = 5, 6, \cdots, 12 \end{array}\right\}.$$

Similarly, if $x_1 \neq 0, x_5 = 0, x_9 \neq 0$, then the chance function of the event $\mathcal{E}$ is

$$f(\boldsymbol{x}) = \Pr\left\{\begin{array}{l} x_1 + x_5 + x_9 = c_1 \\ x_1 + x_2 + x_3 + x_4 \le \xi_1 \\ x_9 + x_{10} + x_{11} + x_{12} \le \xi_3 \\ x_i \ge 0, i = 1, 2, 3, 4, 9, 10, 11, 12 \end{array}\right\}.$$

If $x_1 \neq 0, x_5 \neq 0, x_9 = 0$, then the chance function of the event $\mathcal{E}$ is

$$f(\boldsymbol{x}) = \Pr\left\{\begin{array}{l} x_1 + x_5 + x_9 = c_1 \\ x_1 + x_2 + x_3 + x_4 \le \xi_1 \\ x_5 + x_6 + x_7 + x_8 \le \xi_2 \\ x_i \ge 0, i = 1, 2, \cdots, 8 \end{array}\right\}.$$

If $x_1 = 0, x_5 = 0, x_9 \neq 0$, then the chance function of the event $\mathcal{E}$ is

$$f(\boldsymbol{x}) = \Pr\left\{\begin{array}{l} x_1 + x_5 + x_9 = c_1 \\ x_9 + x_{10} + x_{11} + x_{12} \le \xi_3 \\ x_i \ge 0, i = 9, 10, \cdots, 12 \end{array}\right\}.$$

If $x_1 = 0, x_5 \neq 0, x_9 = 0$, then the chance function of the event $\mathcal{E}$ is

$$f(\boldsymbol{x}) = \Pr\left\{\begin{array}{l} x_1 + x_5 + x_9 = c_1 \\ x_5 + x_6 + x_7 + x_8 \le \xi_2 \\ x_i \ge 0, i = 5, 6, \cdots, 8 \end{array}\right\}.$$

If $x_1 \neq 0, x_5 = 0, x_9 = 0$, then the chance function of the event $\mathcal{E}$ is

$$f(\boldsymbol{x}) = \Pr\left\{\begin{array}{l} x_1 + x_5 + x_9 = c_1 \\ x_1 + x_2 + x_3 + x_4 \le \xi_1 \\ x_i \ge 0, i = 1, 2, \cdots, 4 \end{array}\right\}.$$

Note that the case $x_1 = x_5 = x_9 = 0$ is impossible because $c_1 \neq 0$. It follows that DCP (4.52) is equivalent to the unconstrained model "$\max f(\boldsymbol{x})$".

Dependent-Chance Multiobjective Programming

Since a complex decision system usually undertakes multiple events, there undoubtedly exist multiple potential objectives (some of them are chance functions) in a decision process. A typical formulation of dependent-chance multiobjective programming (DCMOP) is represented as maximizing multiple chance functions subject to an uncertain environment,

$$\left\{\begin{array}{l} \max \left[\begin{array}{l} \Pr\{h_{1k}(\boldsymbol{x}, \boldsymbol{\xi}) \le 0, k = 1, 2, \cdots, q_1\} \\ \Pr\{h_{2k}(\boldsymbol{x}, \boldsymbol{\xi}) \le 0, k = 1, 2, \cdots, q_2\} \\ \cdots \\ \Pr\{h_{mk}(\boldsymbol{x}, \boldsymbol{\xi}) \le 0, k = 1, 2, \cdots, q_m\} \end{array}\right] \\ \text{subject to:} \\ \qquad g_j(\boldsymbol{x}, \boldsymbol{\xi}) \le 0, \quad j = 1, 2, \cdots, p \end{array}\right. \tag{4.53}$$

where $h_{ik}(\boldsymbol{x}, \boldsymbol{\xi}) \le 0, k = 1, 2, \cdots, q_i$ represent events $\mathcal{E}_i$ for $i = 1, 2, \cdots, m$, respectively.

It follows from the principle of uncertainty that we can construct a relationship between decision vectors and chance functions, thus calculating the chance functions by stochastic simulations or traditional methods. Then we can solve DCMOP by utility theory if complete information of the preference function is given by the decision-maker or search for all of the efficient solutions if no information is available. In practice, the decision-maker can provide only partial information. In this case, we have to employ the interactive methods.

Dependent-Chance Goal Programming

When some management targets are given, the objective function may minimize the deviations, positive, negative, or both, with a certain priority structure set by the decision-maker. Then we may formulate the stochastic decision system as the following dependent-chance goal programming (DCGP),

$$\begin{cases} \min \sum\limits_{j=1}^{l} P_j \sum\limits_{i=1}^{m} (u_{ij} d_i^+ \vee 0 + v_{ij} d_i^- \vee 0) \\ \text{subject to:} \\ \quad \Pr\{h_{ik}(\boldsymbol{x}, \boldsymbol{\xi}) \le 0, k = 1, 2, \cdots, q_i\} - b_i = d_i^+, \; i = 1, 2, \cdots, m \\ \quad b_i - \Pr\{h_{ik}(\boldsymbol{x}, \boldsymbol{\xi}) \le 0, k = 1, 2, \cdots, q_i\} = d_i^-, \; i = 1, 2, \cdots, m \\ \quad g_j(\boldsymbol{x}, \boldsymbol{\xi}) \le 0, \qquad j = 1, 2, \cdots, p \end{cases}$$

where P_j is the preemptive priority factor, u_{ij} is the weighting factor corresponding to positive deviation for goal i with priority j assigned, v_{ij} is the weighting factor corresponding to negative deviation for goal i with priority j assigned, $d_i^+ \vee 0$ is the positive deviation from the target of goal i, $d_i^- \vee 0$ is the negative deviation from the target of goal i, b_i is the target value according to goal i, l is the number of priorities, and m is the number of goal constraints.

4.5 Hybrid Intelligent Algorithm

From the mathematical viewpoint, there is no difference between deterministic mathematical programming and stochastic programming except for the fact that there exist uncertain functions in the latter. If the uncertain functions can be converted to their deterministic forms, then we can obtain equivalent deterministic models. However, generally speaking, we cannot do so. It is thus more convenient to deal with them by stochastic simulations. Essentially, there are three types of uncertain functions in stochastic programming as follows:

$$\begin{aligned} U_1 &: \boldsymbol{x} \to E[f(\boldsymbol{x}, \boldsymbol{\xi})], \\ U_2 &: \boldsymbol{x} \to \Pr\{g_j(\boldsymbol{x}, \boldsymbol{\xi}) \le 0, j = 1, 2, \cdots, p\}, \\ U_3 &: \boldsymbol{x} \to \max\left\{\overline{f} \mid \Pr\left\{f(\boldsymbol{x}, \boldsymbol{\xi}) \ge \overline{f}\right\} \ge \alpha\right\}. \end{aligned} \tag{4.54}$$

Stochastic Simulation for $U_1(\boldsymbol{x})$

In order to compute the uncertain function $U_1(\boldsymbol{x})$, we generate ω_k from the probability space $(\Omega, \mathcal{A}, \Pr)$ and produce $\boldsymbol{\xi}_k = \boldsymbol{\xi}(\omega_k)$ for $k = 1, 2, \cdots, N$. *Equivalently, we generate random vectors $\boldsymbol{\xi}_k$ according to the probability distribution Φ for* $k = 1, 2, \cdots, N$. It follows from the strong law of large numbers that

$$\frac{\sum_{k=1}^{N} f(\boldsymbol{x}, \boldsymbol{\xi}_k)}{N} \longrightarrow U_1(\boldsymbol{x}), \quad \text{a.s.} \tag{4.55}$$

as $N \to \infty$. Therefore, the value $U_1(\boldsymbol{x})$ is estimated by

$$\frac{1}{N}\sum_{k=1}^{N} f(\boldsymbol{x}, \boldsymbol{\xi}_k)$$

provided that N is sufficiently large.

Algorithm 4.4 (Stochastic Simulation for $U_1(\boldsymbol{x})$)
Step 1. Set $e = 0$.
Step 2. Generate ω from the probability space $(\Omega, \mathcal{A}, \Pr)$ and produce $\boldsymbol{\xi} = \boldsymbol{\xi}(\omega)$. Equivalently, generate a random vector $\boldsymbol{\xi}$ according to its probability distribution.
Step 3. $e \leftarrow e + f(\boldsymbol{x}, \boldsymbol{\xi})$.
Step 4. Repeat the second and third steps N times.
Step 5. $U_1(\boldsymbol{x}) = e/N$.

Stochastic Simulation for $U_2(\boldsymbol{x})$

In order to compute the uncertain function $U_2(\boldsymbol{x})$, we generate ω_k from the probability space $(\Omega, \mathcal{A}, \Pr)$ and produce $\boldsymbol{\xi}_k = \boldsymbol{\xi}(\omega_k)$ for $k = 1, 2, \cdots, N$. *Equivalently, we generate random vectors $\boldsymbol{\xi}_k$ according to the probability distribution Φ for* $k = 1, 2, \cdots, N$. Let N' denote the number of occasions on which $g_j(\boldsymbol{x}, \boldsymbol{\xi}_k) \le 0, j = 1, 2, \cdots, p$ for $k = 1, 2, \cdots, N$ (i.e., the number of random vectors satisfying the system of inequalities). Let us define

$$h(\boldsymbol{x}, \boldsymbol{\xi}_k) = \begin{cases} 1, \text{ if } g_j(\boldsymbol{x}, \boldsymbol{\xi}_k) \le 0, j = 1, 2, \cdots, p \\ 0, \text{ otherwise.} \end{cases}$$

Then we have $E[h(\boldsymbol{x}, \boldsymbol{\xi}_k)] = U_2(\boldsymbol{x})$ for all k, and $N' = \sum_{k=1}^{N} h(\boldsymbol{x}, \boldsymbol{\xi}_k)$. It follows from the strong law of large numbers that

$$\frac{N'}{N} = \frac{\sum_{k=1}^{N} h(\boldsymbol{x}, \boldsymbol{\xi}_k)}{N}$$

converges a.s. to $U_2(\boldsymbol{x})$. Thus $U_2(\boldsymbol{x})$ can be estimated by N'/N provided that N is sufficiently large.

Algorithm 4.5 (Stochastic Simulation for $U_2(\boldsymbol{x})$)
Step 1. Set $N' = 0$.
Step 2. Generate ω from the probability space $(\Omega, \mathcal{A}, \Pr)$ and produce $\boldsymbol{\xi} = \boldsymbol{\xi}(\omega)$. Equivalently, generate a random vector $\boldsymbol{\xi}$ according to its probability distribution.
Step 3. If $g_j(\boldsymbol{x}, \boldsymbol{\xi}) \le 0$ for $j = 1, 2, \cdots, p$, then $N' \leftarrow N' + 1$.
Step 4. Repeat the second and third steps N times.
Step 5. $U_2(\boldsymbol{x}) = N'/N$.

Stochastic Simulation for $U_3(\boldsymbol{x})$

In order to compute the uncertain function $U_3(\boldsymbol{x})$, we generate ω_k from the probability space $(\Omega, \mathcal{A}, \Pr)$ and produce $\boldsymbol{\xi}_k = \boldsymbol{\xi}(\omega_k)$ for $k = 1, 2, \cdots, N$. *Equivalently, we generate random vectors $\boldsymbol{\xi}_k$ according to the probability distribution Φ for $k = 1, 2, \cdots, N$.* Now we define

$$h(\boldsymbol{x}, \boldsymbol{\xi}_k) = \begin{cases} 1, & \text{if } f(\boldsymbol{x}, \boldsymbol{\xi}_k) \ge \overline{f} \\ 0, & \text{otherwise} \end{cases}$$

for $k = 1, 2, \cdots, N$, which are random variables, and $E[h(\boldsymbol{x}, \boldsymbol{\xi}_k)] = \alpha$ for all k. By the strong law of large numbers, we obtain

$$\frac{\sum_{k=1}^{N} h(\boldsymbol{x}, \boldsymbol{\xi}_k)}{N} \longrightarrow \alpha, \quad \text{a.s.}$$

as $N \to \infty$. Note that the sum $\sum_{k=1}^{N} h(\boldsymbol{x}, \boldsymbol{\xi}_k)$ is just the number of $\boldsymbol{\xi}_k$ satisfying $f(\boldsymbol{x}, \boldsymbol{\xi}_k) \ge \overline{f}$ for $k = 1, 2, \cdots, N$. Thus $\overline{f}$ is just the N'th largest element in the sequence $\{f(\boldsymbol{x}, \boldsymbol{\xi}_1), f(\boldsymbol{x}, \boldsymbol{\xi}_2), \cdots, f(\boldsymbol{x}, \boldsymbol{\xi}_N)\}$, where N' is the integer part of αN.

Algorithm 4.6 (Stochastic Simulation for $U_3(\boldsymbol{x})$)
Step 1. Generate ω_k from the probability space $(\Omega, \mathcal{A}, \Pr)$ and produce $\boldsymbol{\xi}_k = \boldsymbol{\xi}(\omega_k)$ for $k = 1, 2, \cdots, N$. Equivalently, generate random vectors $\boldsymbol{\xi}_k$ according to the probability distribution for $k = 1, 2, \cdots, N$.
Step 2. Set $f_i = f(\boldsymbol{x}, \boldsymbol{\xi}_k)$ for $k = 1, 2, \cdots, N$.
Step 3. Set N' as the integer part of βN.
Step 4. Return the N'th largest element in $\{f_1, f_2, \cdots, f_N\}$ as $U_3(\boldsymbol{x})$.

Neural Network for Approximating Uncertain Functions

Although stochastic simulations are able to compute the uncertain functions, we need relatively simple functions to approximate the uncertain functions because the stochastic simulations are a time-consuming process. In order to speed up the solution process, neural network (NN) is employed to approximate uncertain functions due to the following reasons: (i) NN has the ability to approximate the uncertain functions by using the training data; (ii) NN can compensate for the error of training data (all input-output data obtained by stochastic simulation are clearly not precise); and (iii) NN has the high speed of operation after they are trained.

Hybrid Intelligent Algorithm

Liu [181] integrated stochastic simulation, NN and GA to produce a hybrid intelligent algorithm for solving stochastic programming models.

Algorithm 4.7 (Hybrid Intelligent Algorithm)

Step 1. Generate training input-output data for uncertain functions like

$$\begin{aligned} U_1 &: \boldsymbol{x} \to E[f(\boldsymbol{x}, \boldsymbol{\xi})], \\ U_2 &: \boldsymbol{x} \to \Pr\{g_j(\boldsymbol{x}, \boldsymbol{\xi}) \leq 0, j = 1, 2, \cdots, p\}, \\ U_3 &: \boldsymbol{x} \to \max\left\{\overline{f} \mid \Pr\left\{f(\boldsymbol{x}, \boldsymbol{\xi}) \geq \overline{f}\right\} \geq \alpha\right\} \end{aligned}$$

by the stochastic simulation.

Step 2. Train a neural network to approximate the uncertain functions according to the generated training input-output data.

Step 3. Initialize *pop_size* chromosomes whose feasibility may be checked by the trained neural network.

Step 4. Update the chromosomes by crossover and mutation operations in which the feasibility of offspring may be checked by the trained neural network.

Step 5. Calculate the objective values for all chromosomes by the trained neural network.

Step 6. Compute the fitness of each chromosome according to the objective values.

Step 7. Select the chromosomes by spinning the roulette wheel.

Step 8. Repeat the fourth to seventh steps for a given number of cycles.

Step 9. Report the best chromosome as the optimal solution.

4.6 Numerical Experiments

In order to illustrate its effectiveness, a set of numerical examples has been done, and the results are successful. Here we give some numerical examples which are all performed on a personal computer with the following parameters: the population size is 30, the probability of crossover P_c is 0.3, the probability of mutation P_m is 0.2, and the parameter a in the rank-based evaluation function is 0.05.

Example 4.12. Now we consider the following EVM,

$$\begin{cases} \min E\left[\sqrt{(x_1-\xi_1)^2+(x_2-\xi_2)^2+(x_3-\xi_3)^2}\right] \\ \text{subject to:} \\ \qquad x_1^2+x_2^2+x_3^2 \le 10 \end{cases}$$

where ξ_1 is a uniformly distributed random variable $\mathcal{U}(1,2)$, ξ_2 is a normally distributed random variable $\mathcal{N}(3,1)$, and ξ_3 is an exponentially distributed random variable $\mathcal{EXP}(4)$.

In order to solve this model, we generate input-output data for the uncertain function

$$U:(x_1,x_2,x_3)\to E\left[\sqrt{(x_1-\xi_1)^2+(x_2-\xi_2)^2+(x_3-\xi_3)^2}\right]$$

by stochastic simulation. Then we train an NN (3 input neurons, 5 hidden neurons, 1 output neuron) to approximate the uncertain function U. After that, the trained NN is embedded into a GA to produce a hybrid intelligent algorithm.

A run of the hybrid intelligent algorithm (3000 cycles in simulation, 2000 data in NN, 300 generations in GA) shows that the optimal solution is

$$(x_1^*,x_2^*,x_3^*)=(1.1035,2.1693,2.0191)$$

whose objective value is 3.56.

Example 4.13. Let us consider the following CCP in which there are three decision variables and nine stochastic parameters,

$$\begin{cases} \max \overline{f} \\ \text{subject to:} \\ \qquad \Pr\left\{\xi_1x_1+\xi_2x_2+\xi_3x_3\ge\overline{f}\right\}\ge 0.90 \\ \qquad \Pr\left\{\eta_1x_1^2+\eta_2x_2^2+\eta_3x_3^2\le 8\right\}\ge 0.80 \\ \qquad \Pr\left\{\tau_1x_1^3+\tau_2x_2^3+\tau_3x_3^3\le 15\right\}\ge 0.85 \\ \qquad x_1,x_2,x_3\ge 0 \end{cases}$$

where ξ_1,η_1, and τ_1 are uniformly distributed random variables $\mathcal{U}(1,2)$, $\mathcal{U}(2,3)$, and $\mathcal{U}(3,4)$, respectively, ξ_2,η_2, and τ_2 are normally distributed random variables $\mathcal{N}(1,1)$, $\mathcal{N}(2,1)$, and $\mathcal{N}(3,1)$, respectively, and ξ_3, η_3, and τ_3 are exponentially distributed random variables $\mathcal{EXP}(1)$, $\mathcal{EXP}(2)$, and $\mathcal{EXP}(3)$, respectively,

We employ stochastic simulation to generate input-output data for the uncertain function $U : \boldsymbol{x} \to (U_1(\boldsymbol{x}), U_2(\boldsymbol{x}), U_3(\boldsymbol{x}))$, where

$$\begin{aligned} U_1(\boldsymbol{x}) &= \max\left\{\overline{f} \mid \Pr\left\{\xi_1 x_1 + \xi_2 x_2 + \xi_3 x_3 \geq \overline{f}\right\} \geq 0.90\right\}, \\ U_2(\boldsymbol{x}) &= \Pr\left\{\eta_1 x_1^2 + \eta_2 x_2^2 + \eta_3 x_3^2 \leq 8\right\}, \\ U_3(\boldsymbol{x}) &= \Pr\left\{\tau_1 x_1^3 + \tau_2 x_2^3 + \tau_3 x_3^3 \leq 15\right\}. \end{aligned}$$

Then we train an NN (3 input neurons, 15 hidden neurons, 3 output neurons) to approximate the uncertain function U. Finally, we integrate the trained NN and GA to produce a hybrid intelligent algorithm.

A run of the hybrid intelligent algorithm (5000 cycles in simulation, 3000 training data in NN, 1000 generations in GA) shows that the optimal solution is

$$(x_1^*, x_2^*, x_3^*) = (1.458,\ 0.490,\ 0.811)$$

with objective value $\overline{f}^* = 2.27$.

Example 4.14. Let us now turn our attention to the following DCGP,

$$\begin{cases} \text{lexmin}\left\{d_1^- \vee 0, d_2^- \vee 0, d_3^- \vee 0\right\} \\ \text{subject to:} \\ \qquad 0.92 - \Pr\{x_1 + x_4^2 = 4\} = d_1^- \\ \qquad 0.85 - \Pr\{x_2^2 + x_6 = 3\} = d_2^- \\ \qquad 0.85 - \Pr\{x_3^2 + x_5^2 + x_7^2 = 2\} = d_3^- \\ \qquad x_1 + x_2 + x_3 \leq \xi_1 \\ \qquad x_4 + x_5 \leq \xi_2 \\ \qquad x_6 + x_7 \leq \xi_3 \\ \qquad x_i \geq 0, \quad i = 1, 2, \cdots, 7 \end{cases}$$

where ξ_1, ξ_2, and ξ_3 are uniformly distributed random variable $\mathcal{U}[3,5]$, normally distributed random variable $\mathcal{N}(3.5,1)$, and exponentially distributed random variable $\mathcal{EXP}(9)$, respectively.

In the first priority level, there is only one event $\mathcal{E}_1$ which will be fulfilled by $x_1 + x_4^2 = 4$. It is clear that the support $\mathcal{E}_1^* = \{x_1, x_4\}$ and the dependent support $\mathcal{E}_1^{**} = \{x_1, x_2, x_3, x_4, x_5\}$. Thus the dependent constraints of $\mathcal{E}_1$ are

$$x_1 + x_2 + x_3 \leq \xi_1, \quad x_4 + x_5 \leq \xi_2, \quad x_1, x_2, x_3, x_4, x_5 \geq 0.$$

It follows from the principle of uncertainty that the chance function $f_1(\boldsymbol{x})$ of $\mathcal{E}_1$ is

$$f_1(\boldsymbol{x}) = \Pr\left\{\begin{array}{l} x_1 + x_4^2 = 4 \\ x_1 + x_2 + x_3 \le \xi_1 \\ x_4 + x_5 \le \xi_2 \\ x_1, x_2, x_3, x_4, x_5 \ge 0 \end{array}\right\}.$$

At the second priority level, there is an event $\mathcal{E}_2$ which will be fulfilled by $x_2^2 + x_6 = 3$. The support $\mathcal{E}_2^* = \{x_2, x_6\}$ and the dependent support $\mathcal{E}_2^{**} = \{x_1, x_2, x_3, x_6, x_7\}$. Thus the dependent constraints of $\mathcal{E}_2$ are

$$x_1 + x_2 + x_3 \le \xi_1, \quad x_6 + x_7 \le \xi_3, \quad x_1, x_2, x_3, x_6, x_7 \ge 0.$$

The principle of uncertainty implies that the chance function $f_2(\boldsymbol{x})$ of the event $\mathcal{E}_2$ is

$$f_2(\boldsymbol{x}) = \Pr\left\{\begin{array}{l} x_2^2 + x_6 = 3 \\ x_1 + x_2 + x_3 \le \xi_1 \\ x_6 + x_7 \le \xi_3 \\ x_1, x_2, x_3, x_6, x_7 \ge 0 \end{array}\right\}.$$

At the third priority level, there is an event $\mathcal{E}_3$ which will be fulfilled by $x_3^2 + x_5^2 + x_7^2 = 2$. The support $\mathcal{E}_3^* = \{x_3, x_5, x_7\}$ and the dependent support $\mathcal{E}_3^{**}$ includes all decision variables. Thus all constraints are dependent constraints of $\mathcal{E}_3$. It follows from the principle of uncertainty that the chance function $f_3(\boldsymbol{x})$ of the event $\mathcal{E}_3$ is

$$f_3(\boldsymbol{x}) = \Pr\left\{\begin{array}{l} x_3^2 + x_5^2 + x_7^2 = 2 \\ x_1 + x_2 + x_3 \le \xi_1 \\ x_4 + x_5 \le \xi_2 \\ x_6 + x_7 \le \xi_3 \\ x_1, x_2, x_3, x_4, x_5, x_6, x_7 \ge 0 \end{array}\right\}.$$

We encode a solution into a chromosome $V = (v_1, v_2, v_3, v_4)$, and decode a chromosome into a feasible solution in the following way,

$$\begin{array}{llll} x_1 = v_1, & x_2 = v_2, & x_3 = v_3, & x_4 = \sqrt{4 - v_1}, \\ x_5 = v_4, & x_6 = 3 - v_2^2, & x_7 = \sqrt{2 - v_3^2 - v_4^2}. & \end{array}$$

We first employ stochastic simulation to generate input-output data for the uncertain function $U : (v_1, v_2, v_3, v_4) \to (f_1(\boldsymbol{x}), f_2(\boldsymbol{x}), f_3(\boldsymbol{x}))$. Then we train an NN (4 input neurons, 10 hidden neurons, 3 output neurons) to approximate it. Finally, we embed the trained NN into a GA to produce a hybrid intelligent algorithm.

A run of the hybrid intelligent algorithm (6000 cycles in simulation, 3000 data in NN, 1000 generations in GA) shows that the optimal solution is

$$\boldsymbol{x}^* = (0.1180, 1.7320, 0.1491, 1.9703, 0.0000, 0.0000, 1.4063)$$

which can satisfy the first two goals, but the third objective is 0.05.

Chapter 5
Fuzzy Programming

Fuzzy programming offers a powerful means of handling optimization problems with fuzzy parameters. Fuzzy programming has been used in different ways in the past. Liu and Liu [184] presented a concept of expected value operator of fuzzy variable and provided a spectrum of fuzzy expected value models which optimize the expected objective functions subject to some expected constraints. In addition, Liu and Iwamura [168][169] introduced a spectrum of fuzzy maximax chance-constrained programming, and Liu [171] constructed a spectrum of fuzzy minimax chance-constrained programming in which we assume that the fuzzy constraints will hold with a given credibility level. Liu [172] provided a fuzzy dependent-chance programming theory in order to maximize the chance functions of satisfying some events.

5.1 Fuzzy Variables

The concept of fuzzy set was initialized by Zadeh [325] in 1965. Fuzzy set theory has been well developed and applied in a wide variety of real problems. In order to measure a fuzzy event, Zadeh [328] proposed the concept of possibility measure in 1978. Although possibility measure has been widely used, it is not self-dual. However, a self-dual measure is absolutely needed in both theory and practice. In order to define a self-dual measure, Liu and Liu [184] presented the concept of credibility measure in 2002. In addition, a sufficient and necessary condition for credibility measure was given by Li and Liu [160]. Credibility theory was founded by Liu [186] in 2004 and refined by Liu [189] in 2007 as a branch of mathematics for studying the behavior of fuzzy phenomena.

Let Θ be a nonempty set, and $\mathcal{P}$ the power set of Θ. Each element in $\mathcal{P}$ is called an event. In order to present an axiomatic definition of credibility, it is necessary to assign to each event A a number $\mathrm{Cr}\{A\}$ which indicates the credibility that A will occur. In order to ensure that the number $\mathrm{Cr}\{A\}$ has certain mathematical properties which we intuitively expect a credibility to have, we accept the following four axioms:

B. Liu: Theory and Practice of Uncertain Programming, STUDFUZZ 239, pp. 57–82.
springerlink.com

Axiom 1. *(Normality)* $\text{Cr}\{\Theta\} = 1$.

Axiom 2. *(Monotonicity)* $\text{Cr}\{A\} \le \text{Cr}\{B\}$ *whenever* $A \subset B$.

Axiom 3. *(Self-Duality)* $\text{Cr}\{A\} + \text{Cr}\{A^c\} = 1$ *for any event* A.

Axiom 4. *(Maximality)* $\text{Cr}\{\cup_i A_i\} = \sup_i \text{Cr}\{A_i\}$ *for any events* $\{A_i\}$ *with* $\sup_i \text{Cr}\{A_i\} < 0.5$.

Definition 5.1. *(Liu and Liu [184]) The set function* Cr *is called a credibility measure if it satisfies the normality, monotonicity, self-duality, and maximality axioms.*

Example 5.1. Let $\Theta = \{\theta_1, \theta_2\}$. For this case, there are only four events: $\emptyset, \{\theta_1\}, \{\theta_2\}, \Theta$. Define $\text{Cr}\{\emptyset\} = 0$, $\text{Cr}\{\theta_1\} = 0.7$, $\text{Cr}\{\theta_2\} = 0.3$, and $\text{Cr}\{\Theta\} = 1$. Then the set function Cr is a credibility measure because it satisfies the first four axioms.

Example 5.2. Let Θ be a nonempty set. Define $\text{Cr}\{\emptyset\} = 0$, $\text{Cr}\{\Theta\} = 1$ and $\text{Cr}\{A\} = 1/2$ for any subset A (excluding $\emptyset$ and Θ). Then the set function Cr is a credibility measure.

Example 5.3. Let μ be a nonnegative function on $\Re$ (the set of real numbers) such that $\sup \mu(x) = 1$. Then the set function

$$\text{Cr}\{A\} = \frac{1}{2}\left(\sup_{x \in A} \mu(x) + 1 - \sup_{x \in A^c} \mu(x)\right) \tag{5.1}$$

is a credibility measure on $\Re$.

Theorem 5.1. *Let* Cr *be a credibility measure. Then* $\text{Cr}\{\emptyset\} = 0$ *and* $0 \le \text{Cr}\{A\} \le 1$ *for any* $A \in \mathcal{P}$.

Proof: It follows from Axioms 1 and 3 that $\text{Cr}\{\emptyset\} = 1 - \text{Cr}\{\Theta\} = 1 - 1 = 0$. Since $\emptyset \subset A \subset \Theta$, we have $0 \le \text{Cr}\{A\} \le 1$ by using Axiom 2.

Theorem 5.2. *Let* Θ *be a nonempty set,* $\mathcal{P}$ *the power set of* Θ*, and* Cr *the credibility measure. Then for any* $A, B \in \mathcal{P}$*, we have*

$$\text{Cr}\{A \cup B\} = \text{Cr}\{A\} \vee \text{Cr}\{B\} \quad \text{if } \text{Cr}\{A \cup B\} \le 0.5, \tag{5.2}$$

$$\text{Cr}\{A \cap B\} = \text{Cr}\{A\} \wedge \text{Cr}\{B\} \quad \text{if } \text{Cr}\{A \cap B\} \ge 0.5. \tag{5.3}$$

The above equations hold for not only finite number of events but also infinite number of events.

Proof: If $\text{Cr}\{A \cup B\} < 0.5$, then $\text{Cr}\{A\} \vee \text{Cr}\{B\} < 0.5$ by using Axiom 2. Thus the equation (5.2) follows immediately from Axiom 4. If $\text{Cr}\{A \cup B\} = 0.5$ and (5.2) does not hold, then we have $\text{Cr}\{A\} \vee \text{Cr}\{B\} < 0.5$. It follows from Axiom 4 that

$$\text{Cr}\{A \cup B\} = \text{Cr}\{A\} \vee \text{Cr}\{B\} < 0.5.$$

A contradiction proves (5.2). Next we prove (5.3). Since $\text{Cr}\{A \cap B\} \geq 0.5$, we have $\text{Cr}\{A^c \cup B^c\} \leq 0.5$ by the self-duality. Thus

$$\begin{aligned}\text{Cr}\{A \cap B\} &= 1 - \text{Cr}\{A^c \cup B^c\} = 1 - \text{Cr}\{A^c\} \vee \text{Cr}\{B^c\} \\ &= (1 - \text{Cr}\{A^c\}) \wedge (1 - \text{Cr}\{B^c\}) = \text{Cr}\{A\} \wedge \text{Cr}\{B\}.\end{aligned}$$

The theorem is proved.

Theorem 5.3. *(Liu [186], Credibility Subadditivity Theorem) The credibility measure is subadditive. That is,*

$$\text{Cr}\{A \cup B\} \leq \text{Cr}\{A\} + \text{Cr}\{B\} \tag{5.4}$$

for any events A and B. In fact, credibility measure is not only finitely subadditive but also countably subadditive.

Proof: The argument breaks down into three cases.

Case 1: $\text{Cr}\{A\} < 0.5$ and $\text{Cr}\{B\} < 0.5$. It follows from Axiom 4 that

$$\text{Cr}\{A \cup B\} = \text{Cr}\{A\} \vee \text{Cr}\{B\} \leq \text{Cr}\{A\} + \text{Cr}\{B\}.$$

Case 2: $\text{Cr}\{A\} \geq 0.5$. For this case, by using Axioms 2 and 3, we have $\text{Cr}\{A^c\} \leq 0.5$ and $\text{Cr}\{A \cup B\} \geq \text{Cr}\{A\} \geq 0.5$. Then

$$\begin{aligned}\text{Cr}\{A^c\} &= \text{Cr}\{A^c \cap B\} \vee \text{Cr}\{A^c \cap B^c\} \\ &\leq \text{Cr}\{A^c \cap B\} + \text{Cr}\{A^c \cap B^c\} \\ &\leq \text{Cr}\{B\} + \text{Cr}\{A^c \cap B^c\}.\end{aligned}$$

Applying this inequality, we obtain

$$\begin{aligned}\text{Cr}\{A\} + \text{Cr}\{B\} &= 1 - \text{Cr}\{A^c\} + \text{Cr}\{B\} \\ &\geq 1 - \text{Cr}\{B\} - \text{Cr}\{A^c \cap B^c\} + \text{Cr}\{B\} \\ &= 1 - \text{Cr}\{A^c \cap B^c\} \\ &= \text{Cr}\{A \cup B\}.\end{aligned}$$

Case 3: $\text{Cr}\{B\} \geq 0.5$. This case may be proved by a similar process of Case 2. The theorem is proved.

Definition 5.2. *Let Θ be a nonempty set, $\mathcal{P}$ the power set of Θ, and* Cr *a credibility measure. Then the triplet $(\Theta, \mathcal{P}, \text{Cr})$ is called a credibility space.*

Definition 5.3. *A fuzzy variable is defined as a (measurable) function from a credibility space $(\Theta, \mathcal{P}, \text{Cr})$ to the set of real numbers.*

Definition 5.4. *Let $f : \Re^n \to \Re$ be a function, and $\xi_1, \xi_2, \cdots, \xi_n$ fuzzy variables on the credibility space $(\Theta, \mathcal{P}, \mathrm{Cr})$. Then $\xi = f(\xi_1, \xi_2, \cdots, \xi_n)$ is a fuzzy variable defined as*

$$\xi(\theta) = f(\xi_1(\theta), \xi_2(\theta), \cdots, \xi_n(\theta)) \tag{5.5}$$

for any $\theta \in \Theta$.

Definition 5.5. *An n-dimensional fuzzy vector is defined as a function from a credibility space $(\Theta, \mathcal{P}, \mathrm{Cr})$ to the set of n-dimensional real vectors.*

Theorem 5.4. *The vector $(\xi_1, \xi_2, \cdots, \xi_n)$ is a fuzzy vector if and only if $\xi_1, \xi_2, \cdots, \xi_n$ are fuzzy variables.*

Proof: Write $\boldsymbol{\xi} = (\xi_1, \xi_2, \cdots, \xi_n)$. Suppose that $\boldsymbol{\xi}$ is a fuzzy vector. Then $\xi_1, \xi_2, \cdots, \xi_n$ are functions from Θ to $\Re$. Thus $\xi_1, \xi_2, \cdots, \xi_n$ are fuzzy variables. Conversely, suppose that $\xi_1, \xi_2, \cdots, \xi_n$ are fuzzy variables defined on the credibility space $(\Theta, \mathcal{P}, \mathrm{Cr})$. It is clear that $(\xi_1, \xi_2, \cdots, \xi_n)$ is a function from the credibility space $(\Theta, \mathcal{P}, \mathrm{Cr})$ to $\Re^n$. Hence $(\xi_1, \xi_2, \cdots, \xi_n)$ is a fuzzy vector.

Membership Function

Definition 5.6. *Let ξ be a fuzzy variable defined on the credibility space $(\Theta, \mathcal{P}, \mathrm{Cr})$. Then its membership function is derived from the credibility measure by*

$$\mu(x) = (2\mathrm{Cr}\{\xi = x\}) \wedge 1, \quad x \in \Re. \tag{5.6}$$

Example 5.4. By an *equipossible fuzzy variable* on $[a, b]$ we mean the fuzzy variable whose membership function is given by

$$\mu_1(x) = \begin{cases} 1, \text{ if } a \le x \le b \\ 0, \text{ otherwise.} \end{cases}$$

Example 5.5. By a *triangular fuzzy variable* we mean the fuzzy variable fully determined by the triplet (r_1, r_2, r_3) of crisp numbers with $r_1 < r_2 < r_3$, whose membership function is given by

$$\mu_2(x) = \begin{cases} \dfrac{x - r_1}{r_2 - r_1}, \text{ if } r_1 \le x \le r_2 \\ \dfrac{x - r_3}{r_2 - r_3}, \text{ if } r_2 \le x \le r_3 \\ \quad 0, \qquad \text{otherwise.} \end{cases}$$

Example 5.6. By a *trapezoidal fuzzy variable* we mean the fuzzy variable fully determined by quadruplet (r_1, r_2, r_3, r_4) of crisp numbers with $r_1 < r_2 < r_3 < r_4$, whose membership function is given by

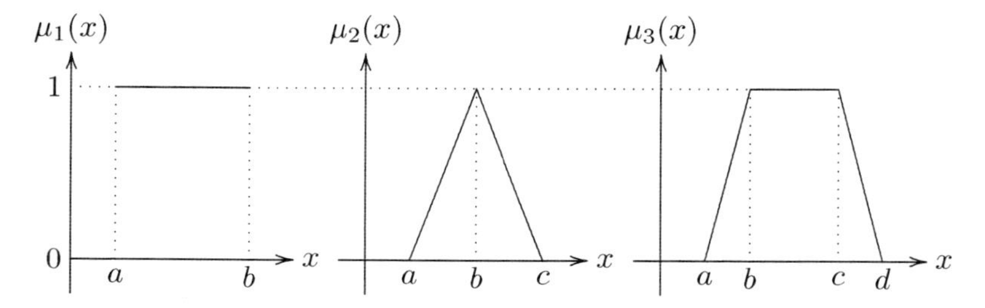

Fig. 5.1 Membership Functions μ_1, μ_2 and μ_3

$$\mu_3(x) = \begin{cases} \dfrac{x - r_1}{r_2 - r_1}, & \text{if } r_1 \le x \le r_2 \\ 1, & \text{if } r_2 \le x \le r_3 \\ \dfrac{x - r_4}{r_3 - r_4}, & \text{if } r_3 \le x \le r_4 \\ 0, & \text{otherwise.} \end{cases}$$

Theorem 5.5. *(Credibility Inversion Theorem) Let ξ be a fuzzy variable with membership function μ. Then for any set B of real numbers, we have*

$$\mathrm{Cr}\{\xi \in B\} = \frac{1}{2}\left(\sup_{x \in B} \mu(x) + 1 - \sup_{x \in B^c} \mu(x)\right). \tag{5.7}$$

Proof: If $\mathrm{Cr}\{\xi \in B\} \le 0.5$, then by Axiom 2, we have $\mathrm{Cr}\{\xi = x\} \le 0.5$ for each $x \in B$. It follows from Axiom 4 that

$$\mathrm{Cr}\{\xi \in B\} = \frac{1}{2}\left(\sup_{x \in B} (2\mathrm{Cr}\{\xi = x\} \wedge 1)\right) = \frac{1}{2} \sup_{x \in B} \mu(x). \tag{5.8}$$

The self-duality of credibility measure implies that $\mathrm{Cr}\{\xi \in B^c\} \ge 0.5$ and $\sup_{x \in B^c} \mathrm{Cr}\{\xi = x\} \ge 0.5$, i.e.,

$$\sup_{x \in B^c} \mu(x) = \sup_{x \in B^c} (2\mathrm{Cr}\{\xi = x\} \wedge 1) = 1. \tag{5.9}$$

It follows from (5.8) and (5.9) that (5.7) holds.

If $\mathrm{Cr}\{\xi \in B\} \ge 0.5$, then $\mathrm{Cr}\{\xi \in B^c\} \le 0.5$. It follows from the first case that

$$\begin{aligned} \mathrm{Cr}\{\xi \in B\} &= 1 - \mathrm{Cr}\{\xi \in B^c\} = 1 - \frac{1}{2}\left(\sup_{x \in B^c} \mu(x) + 1 - \sup_{x \in B} \mu(x)\right) \\ &= \frac{1}{2}\left(\sup_{x \in B} \mu(x) + 1 - \sup_{x \in B^c} \mu(x)\right). \end{aligned}$$

The theorem is proved.

Independence

The independence of fuzzy variables has been discussed by many authors from different angles. Here we use the condition given by Liu and Gao [206].

Definition 5.7. *The fuzzy variables $\xi_1, \xi_2, \cdots, \xi_m$ are said to be independent if*

$$\text{Cr}\left\{\bigcap_{i=1}^{m}\{\xi_i \in B_i\}\right\} = \min_{1 \le i \le m} \text{Cr}\{\xi_i \in B_i\} \tag{5.10}$$

for any sets $B_1, B_2, \cdots, B_m$ of real numbers.

Theorem 5.6. *Let μ_i be membership functions of fuzzy variables ξ_i, $i = 1, 2, \cdots, m$, respectively, and μ the joint membership function of fuzzy vector $(\xi_1, \xi_2, \cdots, \xi_m)$. Then the fuzzy variables $\xi_1, \xi_2, \cdots, \xi_m$ are independent if and only if*

$$\mu(x_1, x_2, \cdots, x_m) = \min_{1 \le i \le m} \mu_i(x_i) \tag{5.11}$$

for any real numbers $x_1, x_2, \cdots, x_m$.

Proof: Suppose that $\xi_1, \xi_2, \cdots, \xi_m$ are independent. It follows that

$$\begin{aligned}
\mu(x_1, x_2, \cdots, x_m) &= \left(2\text{Cr}\left\{\bigcap_{i=1}^{m}\{\xi_i = x_i\}\right\}\right) \wedge 1 \\
&= \left(2 \min_{1 \le i \le m} \text{Cr}\{\xi_i = x_i\}\right) \wedge 1 \\
&= \min_{1 \le i \le m} (2\text{Cr}\{\xi_i = x_i\}) \wedge 1 = \min_{1 \le i \le m} \mu_i(x_i).
\end{aligned}$$

Conversely, for any real numbers $x_1, x_2, \cdots, x_m$ with $\text{Cr}\{\cap_{i=1}^{m}\{\xi_i = x_i\}\} < 0.5$, we have

$$\begin{aligned}
\text{Cr}\left\{\bigcap_{i=1}^{m}\{\xi_i = x_i\}\right\} &= \frac{1}{2}\left(2\text{Cr}\left\{\bigcap_{i=1}^{m}\{\xi_i = x_i\}\right\}\right) \wedge 1 \\
&= \frac{1}{2}\mu(x_1, x_2, \cdots, x_m) = \frac{1}{2} \min_{1 \le i \le m} \mu_i(x_i) \\
&= \frac{1}{2}\left(\min_{1 \le i \le m} (2\text{Cr}\{\xi_i = x_i\}) \wedge 1\right) \\
&= \min_{1 \le i \le m} \text{Cr}\{\xi_i = x_i\}.
\end{aligned}$$

It follows that $\xi_1, \xi_2, \cdots, \xi_m$ are independent. The theorem is proved.

Theorem 5.7. *(Extension Principle of Zadeh) Let $\xi_1, \xi_2, \cdots, \xi_n$ be independent fuzzy variables with membership functions $\mu_1, \mu_2, \cdots, \mu_n$, respectively, and $f : \Re^n \to \Re$ a function. Then the membership function μ of $\xi = f(\xi_1, \xi_2, \cdots, \xi_n)$ is derived from the membership functions $\mu_1, \mu_2, \cdots, \mu_n$ by*

$$\mu(x) = \sup_{x=f(x_1,x_2,\cdots,x_n)} \min_{1\le i\le n} \mu_i(x_i). \tag{5.12}$$

Proof: It follows from Definition 5.6 that the membership function of $\xi = f(\xi_1, \xi_2, \cdots, \xi_n)$ is

$$\begin{aligned}
\mu(x) &= (2\text{Cr}\{f(\xi_1, \xi_2, \cdots, \xi_n) = x\}) \wedge 1 \\
&= \left(2\text{Cr}\left\{\bigcup_{x=f(x_1,x_2,\cdots,x_n)} \{\xi_1 = x_1, \xi_2 = x_2, \cdots, \xi_n = x_n\}\right\}\right) \wedge 1 \\
&= \left(2 \sup_{x=f(x_1,x_2,\cdots,x_n)} \text{Cr}\{\xi_1 = x_1, \xi_2 = x_2, \cdots, \xi_n = x_n\}\right) \wedge 1 \\
&= \left(2 \sup_{x=f(x_1,x_2,\cdots,x_n)} \min_{1\le k\le n} \text{Cr}\{\xi_i = x_i\}\right) \wedge 1 \quad \text{(by independence)} \\
&= \sup_{x=f(x_1,x_2,\cdots,x_n)} \min_{1\le k\le n} (2\text{Cr}\{\xi_i = x_i\}) \wedge 1 \\
&= \sup_{x=f(x_1,x_2,\cdots,x_n)} \min_{1\le i\le n} \mu_i(x_i).
\end{aligned}$$

The theorem is proved.

Example 5.7. By using Theorem 5.7, we may verify that the sum of independent equipossible fuzzy variables $\xi = (a_1, a_2)$ and $\eta = (b_1, b_2)$ is also an equipossible fuzzy variable, and

$$\xi + \eta = (a_1 + b_1, a_2 + b_2).$$

Their product is also an equipossible fuzzy variable, and

$$\xi \cdot \eta = \left(\min_{a_1\le x\le a_2, b_1\le y\le b_2} xy, \max_{a_1\le x\le a_2, b_1\le y\le b_2} xy\right).$$

Example 5.8. The sum of independent triangular fuzzy variables $\xi = (a_1, a_2, a_3)$ and $\eta = (b_1, b_2, b_3)$ is also a triangular fuzzy variable, and

$$\xi + \eta = (a_1 + b_1, a_2 + b_2, a_3 + b_3).$$

The product of a triangular fuzzy variable $\xi = (a_1, a_2, a_3)$ and a scalar number λ is

$$\lambda \cdot \xi = \begin{cases} (\lambda a_1, \lambda a_2, \lambda a_3), \text{ if } \lambda \geq 0 \\ (\lambda a_3, \lambda a_2, \lambda a_1), \text{ if } \lambda < 0. \end{cases}$$

That is, the product of a triangular fuzzy variable and a scalar number is also a triangular fuzzy variable. However, the product of two triangular fuzzy variables is not a triangular one.

Example 5.9. The sum of independent trapezoidal fuzzy variables $\xi = (a_1, a_2, a_3, a_4)$ and $\eta = (b_1, b_2, b_3, b_4)$ is also a trapezoidal fuzzy variable, and $\xi + \eta = (a_1 + b_1, a_2 + b_2, a_3 + b_3, a_4 + b_4)$. The product of a trapezoidal fuzzy variable $\xi = (a_1, a_2, a_3, a_4)$ and a scalar number λ is

$$\lambda \cdot \xi = \begin{cases} (\lambda a_1, \lambda a_2, \lambda a_3, \lambda a_4), \text{ if } \lambda \geq 0 \\ (\lambda a_4, \lambda a_3, \lambda a_2, \lambda a_1), \text{ if } \lambda < 0. \end{cases}$$

That is, the product of a trapezoidal fuzzy variable and a scalar number is also a trapezoidal fuzzy variable.

Example 5.10. Let $\xi_1, \xi_2, \cdots, \xi_n$ be independent fuzzy variables with membership functions $\mu_1, \mu_2, \cdots, \mu_n$, respectively, and $f : \Re^n \to \Re$ a function. Then for any set B of real numbers, the credibility $\text{Cr}\{f(\xi_1, \xi_2, \cdots, \xi_n) \in B\}$ is

$$\frac{1}{2}\left(\sup_{f(x_1,x_2,\cdots,x_n)\in B} \min_{1\leq i\leq n} \mu_i(x_i) + 1 - \sup_{f(x_1,x_2,\cdots,x_n)\in B^c} \min_{1\leq i\leq n} \mu_i(x_i)\right).$$

Expected Value

For fuzzy variables, there are many ways to define an expected value operator. The most general definition of expected value operator of fuzzy variable was given by Liu and Liu [184]. This definition is not only applicable to continuous fuzzy variables but also discrete ones.

Definition 5.8. *(Liu and Liu [184]) Let ξ be a fuzzy variable. Then the expected value of ξ is defined by*

$$E[\xi] = \int_0^{+\infty} \text{Cr}\{\xi \geq r\}\text{d}r - \int_{-\infty}^0 \text{Cr}\{\xi \leq r\}\text{d}r \tag{5.13}$$

provided that at least one of the two integrals is finite.

Example 5.11. Let ξ be the equipossible fuzzy variable (a, b). If $a \geq 0$, then $\text{Cr}\{\xi \leq r\} \equiv 0$ when $r < 0$, and

$$\text{Cr}\{\xi \geq r\} = \begin{cases} 1, & \text{if } r \leq a \\ 0.5, & \text{if } a < r \leq b \\ 0, & \text{if } r > b, \end{cases}$$

$$E[\xi] = \left(\int_0^a 1\mathrm{d}r + \int_a^b 0.5\mathrm{d}r + \int_b^{+\infty} 0\mathrm{d}r\right) - \int_{-\infty}^0 0\mathrm{d}r = \frac{a+b}{2}.$$

If $b \le 0$, then $\mathrm{Cr}\{\xi \ge r\} \equiv 0$ when $r > 0$, and

$$\mathrm{Cr}\{\xi \le r\} = \begin{cases} 1, & \text{if } r \ge b \\ 0.5, & \text{if } a \le r < b \\ 0, & \text{if } r < a, \end{cases}$$

$$E[\xi] = \int_0^{+\infty} 0\mathrm{d}r - \left(\int_{-\infty}^a 0\mathrm{d}r + \int_a^b 0.5\mathrm{d}r + \int_b^0 1\mathrm{d}r\right) = \frac{a+b}{2}.$$

If $a < 0 < b$, then

$$\mathrm{Cr}\{\xi \ge r\} = \begin{cases} 0.5, & \text{if } 0 \le r \le b \\ 0, & \text{if } r > b, \end{cases}$$

$$\mathrm{Cr}\{\xi \le r\} = \begin{cases} 0, & \text{if } r < a \\ 0.5, & \text{if } a \le r \le 0, \end{cases}$$

$$E[\xi] = \left(\int_0^b 0.5\mathrm{d}r + \int_b^{+\infty} 0\mathrm{d}r\right) - \left(\int_{-\infty}^a 0\mathrm{d}r + \int_a^0 0.5\mathrm{d}r\right) = \frac{a+b}{2}.$$

Thus we always have the expected value $(a+b)/2$.

Example 5.12. The triangular fuzzy variable $\xi = (a, b, c)$ has an expected value $E[\xi] = (a + 2b + c)/4$.

Example 5.13. The trapezoidal fuzzy variable $\xi = (a, b, c, d)$ has an expected value $E[\xi] = (a + b + c + d)/4$.

Example 5.14. Let ξ be a continuous nonnegative fuzzy variable with membership function μ. If μ is decreasing on $[0, +\infty)$, then $\mathrm{Cr}\{\xi \ge x\} = \mu(x)/2$ for any $x > 0$, and

$$E[\xi] = \frac{1}{2}\int_0^{+\infty} \mu(x)\mathrm{d}x.$$

Example 5.15. Let ξ be a continuous fuzzy variable with membership function μ. If its expected value exists, and there is a point x_0 such that $\mu(x)$ is increasing on $(-\infty, x_0)$ and decreasing on $(x_0, +\infty)$, then

$$E[\xi] = x_0 + \frac{1}{2}\int_{x_0}^{+\infty} \mu(x)\mathrm{d}x - \frac{1}{2}\int_{-\infty}^{x_0} \mu(x)\mathrm{d}x.$$

Example 5.16. The definition of expected value operator is also applicable to discrete case. Assume that ξ is a simple fuzzy variable whose membership function is given by

$$\mu(x) = \begin{cases} \mu_1, & \text{if } x = x_1 \\ \mu_2, & \text{if } x = x_2 \\ \cdots \\ \mu_m, & \text{if } x = x_m \end{cases} \tag{5.14}$$

where $x_1, x_2, \cdots, x_m$ are distinct numbers. Note that $\mu_1 \vee \mu_2 \vee \cdots \vee \mu_m = 1$. Definition 5.8 implies that the expected value of ξ is

$$E[\xi] = \sum_{i=1}^{m} w_i x_i \tag{5.15}$$

where the weights are given by

$$w_i = \frac{1}{2}\left(\max_{1 \le j \le m} \{\mu_j | x_j \le x_i\} - \max_{1 \le j \le m} \{\mu_j | x_j < x_i\} + \max_{1 \le j \le m} \{\mu_j | x_j \ge x_i\} - \max_{1 \le j \le m} \{\mu_j | x_j > x_i\} \right)$$

for $i = 1, 2, \cdots, m$. It is easy to verify that all $w_i \ge 0$ and the sum of all weights is just 1.

Example 5.17. Consider the fuzzy variable ξ defined by (5.14). Suppose $x_1 < x_2 < \cdots < x_m$. Then the expected value is determined by (5.15) and the weights are given by

$$w_i = \frac{1}{2}\left(\max_{1 \le j \le i} \mu_j - \max_{1 \le j < i} \mu_j + \max_{i \le j \le m} \mu_j - \max_{i < j \le m} \mu_j \right)$$

for $i = 1, 2, \cdots, m$.

Example 5.18. Consider the fuzzy variable ξ defined by (5.14). Suppose $x_1 < x_2 < \cdots < x_m$ and there exists an index k with $1 < k < m$ such that

$$\mu_1 \le \mu_2 \le \cdots \le \mu_k \quad \text{and} \quad \mu_k \ge \mu_{k+1} \ge \cdots \ge \mu_m.$$

Note that $\mu_k \equiv 1$. The expected value $E[\xi]$ is

$$\frac{\mu_1}{2} x_1 + \sum_{i=2}^{k-1} \frac{\mu_i - \mu_{i-1}}{2} x_i + \left(1 - \frac{\mu_{k-1} + \mu_{k+1}}{2}\right) x_k + \sum_{i=k+1}^{m-1} \frac{\mu_i - \mu_{i+1}}{2} x_i + \frac{\mu_m}{2} x_m.$$

Remark 5.1. Let ξ and η be independent fuzzy variables with finite expected values. Then for any numbers a and b, we have

$$E[a\xi + b\eta] = aE[\xi] + bE[\eta]. \tag{5.16}$$

This property is called the linearity of expected value operator of fuzzy variables.

Critical Values

In order to rank fuzzy variables, we may use two critical values: optimistic value and pessimistic value.

Definition 5.9. *(Liu [181]) Let ξ be a fuzzy variable, and $\alpha \in (0,1]$. Then*

$$\xi_{\sup}(\alpha) = \sup\left\{r \mid \mathrm{Cr}\{\xi \geq r\} \geq \alpha\right\} \tag{5.17}$$

is called the α-optimistic value to ξ; and

$$\xi_{\inf}(\alpha) = \inf\left\{r \mid \mathrm{Cr}\{\xi \leq r\} \geq \alpha\right\} \tag{5.18}$$

is called the α-pessimistic value to ξ.

Example 5.19. Let ξ be an equipossible fuzzy variable on $[a,b]$. Then its α-optimistic and α-pessimistic values are

$$\xi_{\sup}(\alpha) = \begin{cases} b, & \text{if } \alpha \leq 0.5 \\ a, & \text{if } \alpha > 0.5, \end{cases} \qquad \xi_{\inf}(\alpha) = \begin{cases} a, & \text{if } \alpha \leq 0.5 \\ b, & \text{if } \alpha > 0.5. \end{cases}$$

Example 5.20. Let $\xi = (r_1, r_2, r_3)$ be a triangular fuzzy variable. Then its α-optimistic and α-pessimistic values are

$$\xi_{\sup}(\alpha) = \begin{cases} 2\alpha r_2 + (1-2\alpha)r_3, & \text{if } \alpha \leq 0.5 \\ (2\alpha-1)r_1 + (2-2\alpha)r_2, & \text{if } \alpha > 0.5, \end{cases}$$

$$\xi_{\inf}(\alpha) = \begin{cases} (1-2\alpha)r_1 + 2\alpha r_2, & \text{if } \alpha \leq 0.5 \\ (2-2\alpha)r_2 + (2\alpha-1)r_3, & \text{if } \alpha > 0.5. \end{cases}$$

Example 5.21. Let $\xi = (r_1, r_2, r_3, r_4)$ be a trapezoidal fuzzy variable. Then its α-optimistic and α-pessimistic values are

$$\xi_{\sup}(\alpha) = \begin{cases} 2\alpha r_3 + (1-2\alpha)r_4, & \text{if } \alpha \leq 0.5 \\ (2\alpha-1)r_1 + (2-2\alpha)r_2, & \text{if } \alpha > 0.5, \end{cases}$$

$$\xi_{\inf}(\alpha) = \begin{cases} (1-2\alpha)r_1 + 2\alpha r_2, & \text{if } \alpha \leq 0.5 \\ (2-2\alpha)r_3 + (2\alpha-1)r_4, & \text{if } \alpha > 0.5. \end{cases}$$

Theorem 5.8. *Let ξ be a fuzzy variable. Then we have*
(a) $\xi_{\inf}(\alpha)$ is an increasing and left-continuous function of α;
(b) $\xi_{\sup}(\alpha)$ is a decreasing and left-continuous function of α.

Proof: (a) It is easy to prove that $\xi_{\inf}(\alpha)$ is an increasing function of α. Next, we prove the left-continuity of $\xi_{\inf}(\alpha)$ with respect to α. Let $\{\alpha_i\}$ be an arbitrary sequence of positive numbers such that $\alpha_i \uparrow \alpha$. Then $\{\xi_{\inf}(\alpha_i)\}$ is an increasing sequence. If the limitation is equal to $\xi_{\inf}(\alpha)$, then the left-continuity is proved. Otherwise, there exists a number z^* such that

$$\lim_{i\to\infty} \xi_{\inf}(\alpha_i) < z^* < \xi_{\inf}(\alpha).$$

Thus $\text{Cr}\{\xi \le z^*\} \ge \alpha_i$ for each i. Letting $i \to \infty$, we get $\text{Cr}\{\xi \le z^*\} \ge \alpha$. Hence $z^* \ge \xi_{\inf}(\alpha)$. A contradiction proves the left-continuity of $\xi_{\inf}(\alpha)$ with respect to α. The part (b) may be proved similarly.

Ranking Criteria

Let ξ and η be two fuzzy variables. Different from the situation of real numbers, there does not exist a natural ordership in a fuzzy world. Thus an important problem appearing in this area is how to rank fuzzy variables. Here we give four ranking criteria.

Expected Value Criterion: We say $\xi > \eta$ if and only if $E[\xi] > E[\eta]$, where E is the expected value operator of fuzzy variable.

Optimistic Value Criterion: We say $\xi > \eta$ if and only if, for some predetermined confidence level $\alpha \in (0, 1]$, we have $\xi_{\sup}(\alpha) > \eta_{\sup}(\alpha)$, where $\xi_{\sup}(\alpha)$ and $\eta_{\sup}(\alpha)$ are the α-optimistic values of ξ and η, respectively.

Pessimistic Value Criterion: We say $\xi > \eta$ if and only if, for some predetermined confidence level $\alpha \in (0, 1]$, we have $\xi_{\inf}(\alpha) > \eta_{\inf}(\alpha)$, where $\xi_{\inf}(\alpha)$ and $\eta_{\inf}(\alpha)$ are the α-pessimistic values of ξ and η, respectively.

Credibility Criterion: We say $\xi > \eta$ if and only if $\text{Cr}\{\xi \ge \overline{r}\} > \text{Cr}\{\eta \ge \overline{r}\}$ for some predetermined level $\overline{r}$.

5.2 Expected Value Model

Assume that $\boldsymbol{x}$ is a decision vector, $\boldsymbol{\xi}$ is a fuzzy vector, $f(\boldsymbol{x}, \boldsymbol{\xi})$ is a return function, and $g_j(\boldsymbol{x}, \boldsymbol{\xi})$ are constraint functions, $j = 1, 2, \cdots, p$. Let us examine the following "fuzzy programming",

$$\begin{cases} \max f(\boldsymbol{x}, \boldsymbol{\xi}) \\ \text{subject to:} \\ \qquad g_j(\boldsymbol{x}, \boldsymbol{\xi}) \le 0, \quad j = 1, 2, \cdots, p. \end{cases} \tag{5.19}$$

Similar to stochastic programming, the model (5.19) is not well-defined because (i) we cannot maximize the fuzzy quantity $f(\boldsymbol{x}, \boldsymbol{\xi})$ (just like that we cannot maximize a random quantity), and (ii) the constraints $g_j(\boldsymbol{x}, \boldsymbol{\xi}) \le 0, j = 1, 2, \cdots, p$ do not produce a crisp feasible set.

Unfortunately, the form of fuzzy programming like (5.19) appears frequently in the literature. Fuzzy programming is a class of mathematical models. Different from fashion or building models, everyone should have the same understanding of the same mathematical model. In other words, a mathematical model must have an unambiguous explanation. The form (5.19) does not have mathematical meaning because it has different interpretations.

In order to obtain the decision with maximum expected return, Liu and Liu [184] provided a spectrum of fuzzy expected value model (EVM),

$$\begin{cases} \max E[f(\boldsymbol{x}, \boldsymbol{\xi})] \\ \text{subject to:} \\ \qquad E[g_j(\boldsymbol{x}, \boldsymbol{\xi})] \le 0, \quad j = 1, 2, \cdots, p \end{cases} \tag{5.20}$$

where $\boldsymbol{x}$ is a decision vector, $\boldsymbol{\xi}$ is a fuzzy vector, $f(\boldsymbol{x}, \boldsymbol{\xi})$ is the return function, and $g_j(\boldsymbol{x}, \boldsymbol{\xi})$ are the constraint functions, $j = 1, 2, \cdots, p$.

In many cases, we may have multiple return functions. Thus we have to employ fuzzy expected value multiobjective programming (EVMOP),

$$\begin{cases} \max\ [E[f_1(\boldsymbol{x}, \boldsymbol{\xi})], E[f_2(\boldsymbol{x}, \boldsymbol{\xi})], \cdots, E[f_m(\boldsymbol{x}, \boldsymbol{\xi})]] \\ \text{subject to:} \\ \qquad E[g_j(\boldsymbol{x}, \boldsymbol{\xi})] \le 0, \quad j = 1, 2, \cdots, p \end{cases} \tag{5.21}$$

where $f_i(\boldsymbol{x}, \boldsymbol{\xi})$ are return functions, $i = 1, 2, \cdots, m$.

In order to balance the multiple conflicting objectives, we may employ the following fuzzy expected value goal programming (EVGP),

$$\begin{cases} \min \sum\limits_{j=1}^{l} P_j \sum\limits_{i=1}^{m} (u_{ij} d_i^+ \vee 0 + v_{ij} d_i^- \vee 0) \\ \text{subject to:} \\ \qquad E[f_i(\boldsymbol{x}, \boldsymbol{\xi})] - b_i = d_i^+, \ i = 1, 2, \cdots, m \\ \qquad b_i - E[f_i(\boldsymbol{x}, \boldsymbol{\xi})] = d_i^-, \ i = 1, 2, \cdots, m \\ \qquad E[g_j(\boldsymbol{x}, \boldsymbol{\xi})] \le 0, \qquad j = 1, 2, \cdots, p \end{cases} \tag{5.22}$$

where P_j is the preemptive priority factor which expresses the relative importance of various goals, $P_j \gg P_{j+1}$, for all j, u_{ij} is the weighting factor corresponding to positive deviation for goal i with priority j assigned, v_{ij} is the weighting factor corresponding to negative deviation for goal i with priority j assigned, $d_i^+ \vee 0$ is the positive deviation from the target of goal i, $d_i^- \vee 0$ is the negative deviation from the target of goal i, f_i is a function in goal constraints, g_j is a function in real constraints, b_i is the target value according to goal i, l is the number of priorities, m is the number of goal constraints, and p is the number of real constraints.

5.3 Chance-Constrained Programming

Assume that $\boldsymbol{x}$ is a decision vector, $\boldsymbol{\xi}$ is a fuzzy vector, $f(\boldsymbol{x}, \boldsymbol{\xi})$ is a return function, and $g_j(\boldsymbol{x}, \boldsymbol{\xi})$ are constraint functions, $j = 1, 2, \cdots, p$. Since the fuzzy constraints $g_j(\boldsymbol{x}, \boldsymbol{\xi}) \le 0, j = 1, 2, \cdots, p$ do not define a deterministic feasible set, a natural idea is to provide a confidence level α at which it is desired that the fuzzy constraints hold. Thus we have a chance constraint as follows,

$$\text{Cr}\left\{g_j(\boldsymbol{x}, \boldsymbol{\xi}) \le 0, j = 1, 2, \cdots, p\right\} \ge \alpha. \tag{5.23}$$

Sometimes, we may employ the following separate chance constraints,

$$\text{Cr}\left\{g_j(\boldsymbol{x}, \boldsymbol{\xi}) \le 0\right\} \ge \alpha_j, \quad j = 1, 2, \cdots, p \tag{5.24}$$

where α_j are confidence levels for $j = 1, 2, \cdots, p$.

Maximax Chance-Constrained Programming

Liu and Iwamura [168][169] suggested a spectrum of fuzzy CCP. When we want to maximize the optimistic return, we have the following fuzzy CCP,

$$\begin{cases} \max\limits_{\boldsymbol{x}} \max\limits_{\overline{f}} \overline{f} \\ \text{subject to:} \\ \qquad \text{Cr}\left\{f(\boldsymbol{x}, \boldsymbol{\xi}) \ge \overline{f}\right\} \ge \beta \\ \qquad \text{Cr}\left\{g_j(\boldsymbol{x}, \boldsymbol{\xi}) \le 0, j = 1, 2, \cdots, p\right\} \ge \alpha \end{cases} \tag{5.25}$$

where α and β are the predetermined confidence levels, and $\max \overline{f}$ is the β-optimistic return.

If there are multiple objectives, then we have a chance-constrained multiobjective programming (CCMOP),

$$\begin{cases} \max\limits_{\boldsymbol{x}} \left[\max\limits_{\overline{f}_1} \overline{f}_1, \max\limits_{\overline{f}_2} \overline{f}_2, \cdots, \max\limits_{\overline{f}_m} \overline{f}_m\right] \\ \text{subject to:} \\ \qquad \text{Cr}\left\{f_i(\boldsymbol{x}, \boldsymbol{\xi}) \ge \overline{f}_i\right\} \ge \beta_i, i = 1, 2, \cdots, m \\ \qquad \text{Cr}\{g_j(\boldsymbol{x}, \boldsymbol{\xi}) \le 0\} \ge \alpha_j, \quad j = 1, 2, \cdots, p \end{cases} \tag{5.26}$$

where $\alpha_1, \alpha_2, \cdots, \alpha_p, \beta_1, \beta_2, \cdots, \beta_m$ are the predetermined confidence levels, and $\max \overline{f}_i$ are the β_i-optimistic values to the return functions $f_i(\boldsymbol{x}, \boldsymbol{\xi})$, $i = 1, 2, \cdots, m$, respectively.

We can also formulate the fuzzy decision system as a minimin chance-constrained goal programming (CCGP) according to the priority structure and target levels set by the decision-maker:

$$\begin{cases} \min\limits_{\boldsymbol{x}} \sum\limits_{j=1}^{l} P_j \sum\limits_{i=1}^{m} \left(u_{ij} \left(\min\limits_{d_i^+} d_i^+ \vee 0 \right) + v_{ij} \left(\min\limits_{d_i^-} d_i^- \vee 0 \right) \right) \\ \text{subject to:} \\ \qquad \text{Cr}\left\{ f_i(\boldsymbol{x}, \boldsymbol{\xi}) - b_i \le d_i^+ \right\} \ge \beta_i^+, \ i = 1, 2, \cdots, m \\ \qquad \text{Cr}\left\{ b_i - f_i(\boldsymbol{x}, \boldsymbol{\xi}) \le d_i^- \right\} \ge \beta_i^-, \ i = 1, 2, \cdots, m \\ \qquad \text{Cr}\left\{ g_j(\boldsymbol{x}, \boldsymbol{\xi}) \le 0 \right\} \ge \alpha_j, \qquad j = 1, 2, \cdots, p \end{cases} \tag{5.27}$$

where P_j is the preemptive priority factor which expresses the relative importance of various goals, $P_j \gg P_{j+1}$, for all j, u_{ij} is the weighting factor corresponding to positive deviation for goal i with priority j assigned, v_{ij} is the weighting factor corresponding to negative deviation for goal i with priority j assigned, $\min d_i^+ \vee 0$ is the β_i^+-optimistic positive deviation from the target of goal i, $\min d_i^- \vee 0$ is the β_i^--optimistic negative deviation from the target of goal i, f_i is a function in goal constraints, g_j is a function in real constraints, b_i is the target value according to goal i, l is the number of priorities, m is the number of goal constraints, p is the number of real constraints.

Minimax Chance-Constrained Programming

In fact, maximax CCP models are essentially a type of optimistic models which maximize the maximum possible return. This section introduces a spectrum of minimax CCP constructed by Liu [171], which will select the alternative that provides the best of the worst possible return.

If we want to maximize the pessimistic return, then we have the following minimax CCP,

$$\begin{cases} \max\limits_{\boldsymbol{x}} \min\limits_{\overline{f}} \overline{f} \\ \text{subject to:} \\ \qquad \text{Cr}\left\{ f(\boldsymbol{x}, \boldsymbol{\xi}) \le \overline{f} \right\} \ge \beta \\ \qquad \text{Cr}\left\{ g_j(\boldsymbol{x}, \boldsymbol{\xi}) \le 0, j = 1, 2, \cdots, p \right\} \ge \alpha \end{cases} \tag{5.28}$$

where $\min \overline{f}$ is the β-pessimistic return.

If there are multiple objectives, we may employ the following minimax CCMOP,

$$\begin{cases} \max\limits_{\boldsymbol{x}} \left[\min\limits_{\overline{f}_1} \overline{f}_1, \ \min\limits_{\overline{f}_2} \overline{f}_2, \ \cdots, \ \min\limits_{\overline{f}_m} \overline{f}_m \right] \\ \text{subject to:} \\ \qquad \text{Cr}\left\{ f_i(\boldsymbol{x}, \boldsymbol{\xi}) \le \overline{f}_i \right\} \ge \beta_i, \ i = 1, 2, \cdots, m \\ \qquad \text{Cr}\left\{ g_j(\boldsymbol{x}, \boldsymbol{\xi}) \le 0 \right\} \ge \alpha_j, \ \ j = 1, 2, \cdots, p \end{cases} \tag{5.29}$$

where α_j and β_i are confidence levels, and $\min \overline{f}_i$ are the β_i-pessimistic values to the return functions $f_i(\boldsymbol{x}, \boldsymbol{\xi})$, $i = 1, 2, \cdots, m$, respectively.

According to the priority structure and target levels set by the decision-maker, the minimax CCGP is written as follows,

$$\begin{cases} \min\limits_{\boldsymbol{x}} \sum\limits_{j=1}^{l} P_j \sum\limits_{i=1}^{m} \left[u_{ij} \left(\max\limits_{d_i^+} d_i^+ \vee 0 \right) + v_{ij} \left(\max\limits_{d_i^-} d_i^- \vee 0 \right) \right] \\ \text{subject to:} \\ \quad \text{Cr}\left\{ f_i(\boldsymbol{x}, \boldsymbol{\xi}) - b_i \geq d_i^+ \right\} \geq \beta_i^+, \ i = 1, 2, \cdots, m \\ \quad \text{Cr}\left\{ b_i - f_i(\boldsymbol{x}, \boldsymbol{\xi}) \geq d_i^- \right\} \geq \beta_i^-, \ i = 1, 2, \cdots, m \\ \quad \text{Cr}\left\{ g_j(\boldsymbol{x}, \boldsymbol{\xi}) \leq 0 \right\} \geq \alpha_j, \qquad j = 1, 2, \cdots, p \end{cases} \tag{5.30}$$

where P_j is the preemptive priority factor which expresses the relative importance of various goals, $P_j \gg P_{j+1}$, for all j, u_{ij} is the weighting factor corresponding to positive deviation for goal i with priority j assigned, v_{ij} is the weighting factor corresponding to negative deviation for goal i with priority j assigned, $\max d_i^+ \vee 0$ is the β_i^+-pessimistic positive deviation from the target of goal i, $\max d_i^- \vee 0$ is the β_i^--pessimistic negative deviation from the target of goal i, b_i is the target value according to goal i, l is the number of priorities, and m is the number of goal constraints.

Crisp Equivalents

One way of solving fuzzy CCP is to convert the chance constraint

$$\text{Cr}\left\{ g(\boldsymbol{x}, \boldsymbol{\xi}) \leq 0 \right\} \geq \alpha \tag{5.31}$$

into its crisp equivalent and then solve the equivalent crisp model by the traditional solution process. Please note that

(a) the system constraints $\text{Cr}\{g_j(\boldsymbol{x}, \boldsymbol{\xi}) \leq 0\} \geq \alpha_j$, $j = 1, 2, \cdots, p$ are a set of form (5.31);
(b) the objective constraint $\text{Cr}\{f(\boldsymbol{x}, \boldsymbol{\xi}) \geq \overline{f}\} \geq \beta$ coincides with the form (5.31) by defining $g(\boldsymbol{x}, \boldsymbol{\xi}) = \overline{f} - f(\boldsymbol{x}, \boldsymbol{\xi})$;
(c) the fuzzy constraint $\text{Cr}\{f(\boldsymbol{x}, \boldsymbol{\xi}) \leq \overline{f}\} \geq \beta$ coincides with the form (5.31) by defining $g(\boldsymbol{x}, \boldsymbol{\xi}) = f(\boldsymbol{x}, \boldsymbol{\xi}) - \overline{f}$;
(d) $\text{Cr}\{b - f(\boldsymbol{x}, \boldsymbol{\xi}) \leq d^-\} \geq \beta$ and $\text{Cr}\{f(\boldsymbol{x}, \boldsymbol{\xi}) - b \leq d^+\} \geq \beta$ coincide with the form (5.31) by defining $g(\boldsymbol{x}, \boldsymbol{\xi}) = b - f(\boldsymbol{x}, \boldsymbol{\xi}) - d^-$ and $g(\boldsymbol{x}, \boldsymbol{\xi}) = f(\boldsymbol{x}, \boldsymbol{\xi}) - b - d^+$, respectively; and
(e) $\text{Cr}\{b - f(\boldsymbol{x}, \boldsymbol{\xi}) \geq d^-\} \geq \beta$ and $\text{Cr}\{f(\boldsymbol{x}, \boldsymbol{\xi}) - b \geq d^+\} \geq \beta$ coincide with the form (5.31) by defining $g(\boldsymbol{x}, \boldsymbol{\xi}) = f(\boldsymbol{x}, \boldsymbol{\xi}) + d^- - b$ and $g(\boldsymbol{x}, \boldsymbol{\xi}) = b - f(\boldsymbol{x}, \boldsymbol{\xi}) + d^+$, respectively.

This section presents some useful results.

Theorem 5.9. *Assume that the fuzzy vector* $\boldsymbol{\xi}$ *degenerates to a fuzzy variable* ξ *with continuous membership function* μ, *and the function* $g(\boldsymbol{x}, \boldsymbol{\xi})$ *has the form* $g(\boldsymbol{x}, \boldsymbol{\xi}) = h(\boldsymbol{x}) - \xi$. *Then* $\mathrm{Cr}\{g(\boldsymbol{x}, \boldsymbol{\xi}) \le 0\} \ge \alpha$ *if and only if* $h(\boldsymbol{x}) \le K_\alpha$, *where*

$$K_\alpha = \begin{cases} \sup\left\{K | K = \mu^{-1}(2\alpha)\right\}, & \text{if } \alpha < 1/2 \\ \inf\left\{K | K = \mu^{-1}(2(1-\alpha))\right\}, & \text{if } \alpha \ge 1/2. \end{cases} \tag{5.32}$$

Proof: It is easy to verify the theorem by the relation between credibility measure and membership function.

Theorem 5.10. *Assume that the function* $g(\boldsymbol{x}, \boldsymbol{\xi})$ *can be rewritten as,*

$$g(\boldsymbol{x}, \boldsymbol{\xi}) = h_1(\boldsymbol{x})\xi_1 + h_2(\boldsymbol{x})\xi_2 + \cdots + h_t(\boldsymbol{x})\xi_t + h_0(\boldsymbol{x})$$

where ξ_k *are trapezoidal fuzzy variables* $(r_{k1}, r_{k2}, r_{k3}, r_{k4})$, $k = 1, 2, \cdots, t$, *respectively. We define two functions* $h_k^+(\boldsymbol{x}) = h_k(\boldsymbol{x}) \vee 0$ *and* $h_k^-(\boldsymbol{x}) = -(h_k(\boldsymbol{x}) \wedge 0)$ *for* $k = 1, 2, \cdots, t$. *Then we have*
(a) when $\alpha < 1/2$, $\mathrm{Cr}\{g(\boldsymbol{x}, \boldsymbol{\xi}) \le 0\} \ge \alpha$ *if and only if*

$$\begin{aligned} &(1-2\alpha)\sum_{k=1}^{t}\left[r_{k1}h_k^+(\boldsymbol{x}) - r_{k4}h_k^-(\boldsymbol{x})\right] \\ &\qquad + 2\alpha\sum_{k=1}^{t}\left[r_{k2}h_k^+(\boldsymbol{x}) - r_{k3}h_k^-(\boldsymbol{x})\right] + h_0(\boldsymbol{x}) \le 0; \end{aligned} \tag{5.33}$$

(b) when $\alpha \ge 1/2$, $\mathrm{Cr}\{g(\boldsymbol{x}, \boldsymbol{\xi}) \le 0\} \ge \alpha$ *if and only if*

$$\begin{aligned} &(2-2\alpha)\sum_{k=1}^{t}\left[r_{k3}h_k^+(\boldsymbol{x}) - r_{k2}h_k^-(\boldsymbol{x})\right] \\ &\qquad + (2\alpha-1)\sum_{k=1}^{t}\left[r_{k4}h_k^+(\boldsymbol{x}) - r_{k1}h_k^-(\boldsymbol{x})\right] + h_0(\boldsymbol{x}) \le 0. \end{aligned} \tag{5.34}$$

Proof: It is clear that the functions $h_k^+(\boldsymbol{x})$ and $h_k^-(\boldsymbol{x})$ are all nonnegative and $h_k(\boldsymbol{x}) = h_k^+(\boldsymbol{x}) - h_k^-(\boldsymbol{x})$. Thus we have

$$\begin{aligned} g(\boldsymbol{x}, \boldsymbol{\xi}) &= \sum_{k=1}^{t} h_k(\boldsymbol{x})\xi_k + h_0(\boldsymbol{x}) \\ &= \sum_{k=1}^{t}\left[h_k^+(\boldsymbol{x}) - h_k^-(\boldsymbol{x})\right]\xi_k + h_0(\boldsymbol{x}) \\ &= \sum_{k=1}^{t}\left[h_k^+(\boldsymbol{x})\xi_k + h_k^-(\boldsymbol{x})\xi_k'\right] + h_0(\boldsymbol{x}) \end{aligned}$$

where ξ_k' are also trapezoidal fuzzy variables,

$$\xi_k' = (-r_{k4}, -r_{k3}, -r_{k2}, -r_{k1}), \quad k = 1, 2, \cdots, t.$$

By the addition and multiplication operations of trapezoidal fuzzy variables, the function $g(\boldsymbol{x}, \boldsymbol{\xi})$ is also a trapezoidal fuzzy variable determined by the quadruple

$$g(\boldsymbol{x}, \boldsymbol{\xi}) = \begin{pmatrix} \sum_{k=1}^{t} \left[r_{k1}h_k^+(\boldsymbol{x}) - r_{k4}h_k^-(\boldsymbol{x})\right] + h_0(\boldsymbol{x}) \\ \sum_{k=1}^{t} \left[r_{k2}h_k^+(\boldsymbol{x}) - r_{k3}h_k^-(\boldsymbol{x})\right] + h_0(\boldsymbol{x}) \\ \sum_{k=1}^{t} \left[r_{k3}h_k^+(\boldsymbol{x}) - r_{k2}h_k^-(\boldsymbol{x})\right] + h_0(\boldsymbol{x}) \\ \sum_{k=1}^{t} \left[r_{k4}h_k^+(\boldsymbol{x}) - r_{k1}h_k^-(\boldsymbol{x})\right] + h_0(\boldsymbol{x}) \end{pmatrix}^T.$$

It follows that the results hold.

5.4 Dependent-Chance Programming

Liu [172] provided a fuzzy dependent-chance programming (DCP) theory in which the underlying philosophy is based on selecting the decision with maximum credibility to meet the event.

Basic Concepts

Uncertain environment, event and chance function are key elements in the framework of DCP in a stochastic environment. Let us redefine them in fuzzy environments.

Definition 5.10. *By uncertain environment (in this case the fuzzy environment) we mean the fuzzy constraints represented by*

$$g_j(\boldsymbol{x}, \boldsymbol{\xi}) \le 0, \quad j = 1, 2, \cdots, p \tag{5.35}$$

where $\boldsymbol{x}$ is a decision vector, and $\boldsymbol{\xi}$ is a fuzzy vector.

Definition 5.11. *By event we mean a system of fuzzy inequalities,*

$$h_k(\boldsymbol{x}, \boldsymbol{\xi}) \le 0, \quad k = 1, 2, \cdots, q \tag{5.36}$$

where $\boldsymbol{x}$ is a decision vector, and $\boldsymbol{\xi}$ is a fuzzy vector.

Definition 5.12. *The chance function of an event $\mathcal{E}$ characterized by (5.36) is defined as the credibility measure of the event $\mathcal{E}$, i.e.,*

$$f(\boldsymbol{x}) = \mathrm{Cr}\{h_k(\boldsymbol{x}, \boldsymbol{\xi}) \le 0, \, k = 1, 2, \cdots, q\} \tag{5.37}$$

subject to the uncertain environment (5.35).

The concepts of the support, dependent support, active constraint, and dependent constraint are the same with those in stochastic case. Thus, for each decision $\boldsymbol{x}$ and realization $\boldsymbol{\xi}$, an event $\mathcal{E}$ is said to be consistent in the uncertain environment if the following two conditions hold: (i) $h_k(\boldsymbol{x}, \boldsymbol{\xi}) \le 0$, $k = 1, 2, \cdots, q$; and (ii) $g_j(\boldsymbol{x}, \boldsymbol{\xi}) \le 0$, $j \in J$, where J is the index set of all dependent constraints. In order to compute the chance function of a fuzzy event, we need the following principle of uncertainty.

Principle of Uncertainty: *The chance of a fuzzy event is the credibility that the event is consistent in the uncertain environment.*

Assume that there are m events $\mathcal{E}_i$ characterized by $h_{ik}(\boldsymbol{x}, \boldsymbol{\xi}) \le 0, k = 1, 2, \cdots, q_i$ for $i = 1, 2, \cdots, m$ in the uncertain environment $g_j(\boldsymbol{x}, \boldsymbol{\xi}) \le 0, j = 1, 2, \cdots, p$. The principle of uncertainty implies that the chance function of the ith event $\mathcal{E}_i$ in the uncertain environment is

$$f_i(\boldsymbol{x}) = \text{Cr}\left\{\begin{array}{l} h_{ik}(\boldsymbol{x}, \boldsymbol{\xi}) \le 0, k = 1, 2, \cdots, q_i \\ g_j(\boldsymbol{x}, \boldsymbol{\xi}) \le 0, j \in J_i \end{array}\right\} \tag{5.38}$$

where J_i are defined by

$$J_i = \left\{ j \in \{1, 2, \cdots, p\} \;\middle|\; g_j(\boldsymbol{x}, \boldsymbol{\xi}) \le 0 \text{ is a dependent constraint of } \mathcal{E}_i \right\}$$

for $i = 1, 2, \cdots, m$.

General Models

A typical formulation of DCP in a fuzzy environment is given as follows:

$$\left\{\begin{array}{l} \max \text{Cr}\{h_k(\boldsymbol{x}, \boldsymbol{\xi}) \le 0, k = 1, 2, \cdots, q\} \\ \text{subject to:} \\ \qquad g_j(\boldsymbol{x}, \boldsymbol{\xi}) \le 0, \quad j = 1, 2, \cdots, p \end{array}\right. \tag{5.39}$$

where $\boldsymbol{x}$ is an n-dimensional decision vector, $\boldsymbol{\xi}$ is a fuzzy vector, the event $\mathcal{E}$ is characterized by $h_k(\boldsymbol{x}, \boldsymbol{\xi}) \le 0, k = 1, 2, \cdots, q$, and the uncertain environment is described by the fuzzy constraints $g_j(\boldsymbol{x}, \boldsymbol{\xi}) \le 0, j = 1, 2, \cdots, p$.

Fuzzy DCP (5.39) reads as "maximizing the credibility of the fuzzy event $h_k(\boldsymbol{x}, \boldsymbol{\xi}) \le 0, k = 1, 2, \cdots, q$ subject to the uncertain environment $g_j(\boldsymbol{x}, \boldsymbol{\xi}) \le 0, j = 1, 2, \cdots, p$".

Since a complex decision system usually undertakes multiple tasks, there undoubtedly exist multiple potential objectives. A typical formulation of fuzzy dependent-chance multiobjective programming (DCMOP) is given as follows,

$$\left\{\begin{array}{l} \max \left[\begin{array}{l} \text{Cr}\{h_{1k}(\boldsymbol{x}, \boldsymbol{\xi}) \le 0, k = 1, 2, \cdots, q_1\} \\ \text{Cr}\{h_{2k}(\boldsymbol{x}, \boldsymbol{\xi}) \le 0, k = 1, 2, \cdots, q_2\} \\ \cdots \\ \text{Cr}\{h_{mk}(\boldsymbol{x}, \boldsymbol{\xi}) \le 0, k = 1, 2, \cdots, q_m\} \end{array}\right] \\ \text{subject to:} \\ \qquad g_j(\boldsymbol{x}, \boldsymbol{\xi}) \le 0, \quad j = 1, 2, \cdots, p \end{array}\right. \tag{5.40}$$

where $h_{ik}(\boldsymbol{x}, \boldsymbol{\xi}) \le 0, k = 1, 2, \cdots, q_i$ represent events $\mathcal{E}_i$ for $i = 1, 2, \cdots, m$, respectively.

Dependent-chance goal programming (DCGP) in fuzzy environment may be considered as an extension of goal programming in a complex fuzzy decision system. When some management targets are given, the objective function may minimize the deviations, positive, negative, or both, with a certain priority structure. Thus we can formulate a fuzzy decision system as a DCGP according to the priority structure and target levels set by the decision-maker,

$$\begin{cases} \min \sum_{j=1}^{l} P_j \sum_{i=1}^{m} (u_{ij} d_i^+ \vee 0 + v_{ij} d_i^- \vee 0) \\ \text{subject to:} \\ \quad \text{Cr}\{h_{ik}(\boldsymbol{x}, \boldsymbol{\xi}) \le 0, k = 1, 2, \cdots, q_i\} - b_i = d_i^+, \; i = 1, 2, \cdots, m \\ \quad b_i - \text{Cr}\{h_{ik}(\boldsymbol{x}, \boldsymbol{\xi}) \le 0, k = 1, 2, \cdots, q_i\} = d_i^-, \; i = 1, 2, \cdots, m \\ \quad g_j(\boldsymbol{x}, \boldsymbol{\xi}) \le 0, \qquad j = 1, 2, \cdots, p \end{cases}$$

where P_j is the preemptive priority factor which expresses the relative importance of various goals, $P_j \gg P_{j+1}$, for all j, u_{ij} is the weighting factor corresponding to positive deviation for goal i with priority j assigned, v_{ij} is the weighting factor corresponding to negative deviation for goal i with priority j assigned, $d_i^+ \vee 0$ is the positive deviation from the target of goal i, $d_i^- \vee 0$ is the negative deviation from the target of goal i, g_j is a function in system constraints, b_i is the target value according to goal i, l is the number of priorities, m is the number of goal constraints, and p is the number of system constraints.

5.5 Hybrid Intelligent Algorithm

In order to solve general fuzzy programming models, we must deal with the following three types of uncertain function:

$$\begin{aligned} &U_1 : \boldsymbol{x} \to E[f(\boldsymbol{x}, \boldsymbol{\xi})], \\ &U_2 : \boldsymbol{x} \to \text{Cr}\{g_j(\boldsymbol{x}, \boldsymbol{\xi}) \le 0, j = 1, 2, \cdots, p\}, \\ &U_3 : \boldsymbol{x} \to \max\left\{\overline{f} \mid \text{Cr}\left\{f(\boldsymbol{x}, \boldsymbol{\xi}) \ge \overline{f}\right\} \ge \alpha\right\}. \end{aligned} \tag{5.41}$$

Fuzzy Simulation for $U_1(\boldsymbol{x})$

In order to compute the expected value $U_1(\boldsymbol{x})$, the following procedure may be used. We randomly generate θ_k from the credibility space $(\Theta, \mathcal{P}, \text{Cr})$, write $\nu_k = (2\text{Cr}\{\theta_k\}) \wedge 1$ and produce $\boldsymbol{\xi}_k = \boldsymbol{\xi}(\theta_k)$, $k = 1, 2, \cdots, N$, respectively. *Equivalently, we randomly generate* $\boldsymbol{\xi}_k$ *and write* $\nu_k = \mu(\boldsymbol{\xi}_k)$ *for* $k = 1, 2, \cdots, N$, *where* μ *is the membership function of* $\boldsymbol{\xi}$. Then for any number $r \ge 0$, the credibility $\text{Cr}\{f(\boldsymbol{x}, \boldsymbol{\xi}) \ge r\}$ can be estimated by

$$\frac{1}{2}\left(\max_{1 \le k \le N}\{\nu_k \mid f(\boldsymbol{x}, \boldsymbol{\xi}_k) \ge r\} + \min_{1 \le k \le N}\{1 - \nu_k \mid f(\boldsymbol{x}, \boldsymbol{\xi}_k) < r\}\right)$$

and for any number $r < 0$, the credibility $\text{Cr}\{f(\boldsymbol{x}, \boldsymbol{\xi}) \leq r\}$ can be estimated by

$$\frac{1}{2}\left(\max_{1\leq k\leq N}\{\nu_k \mid f(\boldsymbol{x}, \boldsymbol{\xi}_k) \leq r\} + \min_{1\leq k\leq N}\{1-\nu_k \mid f(\boldsymbol{x}, \boldsymbol{\xi}_k) > r\}\right)$$

provided that N is sufficiently large. Thus $U_1(\boldsymbol{x})$ may be estimated by the following procedure.

Algorithm 5.1 (Fuzzy Simulation for $U_1(\boldsymbol{x})$)
Step 1. Set $e = 0$.
Step 2. Randomly generate θ_k from the credibility space $(\Theta, \mathcal{P}, \text{Cr})$, write $\nu_k = (2\text{Cr}\{\theta_k\}) \wedge 1$ and produce $\boldsymbol{\xi}_k = \boldsymbol{\xi}(\theta_k)$, $k = 1, 2, \cdots, N$, respectively. *Equivalently, randomly generate* $\boldsymbol{\xi}_k$ *and write* $\nu_k = \mu(\boldsymbol{\xi}_k)$ *for* $k = 1, 2, \cdots, N$, *where* μ *is the membership function of* $\boldsymbol{\xi}$.
Step 3. Set two numbers $a = f(\boldsymbol{x}, \boldsymbol{\xi}_1) \wedge f(\boldsymbol{x}, \boldsymbol{\xi}_2) \wedge \cdots \wedge f(\boldsymbol{x}, \boldsymbol{\xi}_N)$ and $b = f(\boldsymbol{x}, \boldsymbol{\xi}_1) \vee f(\boldsymbol{x}, \boldsymbol{\xi}_2) \vee \cdots \vee f(\boldsymbol{x}, \boldsymbol{\xi}_N)$.
Step 4. Randomly generate r from $[a, b]$.
Step 5. If $r \geq 0$, then $e \leftarrow e + \text{Cr}\{f(\boldsymbol{x}, \boldsymbol{\xi}) \geq r\}$.
Step 6. If $r < 0$, then $e \leftarrow e - \text{Cr}\{f(\boldsymbol{x}, \boldsymbol{\xi}) \leq r\}$.
Step 7. Repeat the fourth to sixth steps for N times.
Step 8. $U_1(\boldsymbol{x}) = a \vee 0 + b \wedge 0 + e \cdot (b - a)/N$.

Fuzzy Simulation for $U_2(\boldsymbol{x})$

In order to compute the uncertain function $U_2(\boldsymbol{x})$, we randomly generate θ_k from the credibility space $(\Theta, \mathcal{P}, \text{Cr})$, write $\nu_k = (2\text{Cr}\{\theta_k\}) \wedge 1$ and produce $\boldsymbol{\xi}_k = \boldsymbol{\xi}(\theta_k)$, $k = 1, 2, \cdots, N$, respectively. *Equivalently, we randomly generate* $\boldsymbol{\xi}_k$ *and write* $\nu_k = \mu(\boldsymbol{\xi}_k)$ *for* $k = 1, 2, \cdots, N$, *where* μ *is the membership function of* $\boldsymbol{\xi}$. Then the credibility $U_2(\boldsymbol{x})$ can be estimated by the formula,

$$\frac{1}{2}\left(\max_{1\leq k\leq N}\left\{\nu_k \,\Big|\, \begin{array}{c} g_j(\boldsymbol{x}, \boldsymbol{\xi}_k) \leq 0 \\ j = 1, 2, \cdots, p \end{array}\right\} + \min_{1\leq k\leq N}\left\{1-\nu_k \,\Big|\, \begin{array}{c} g_j(\boldsymbol{x}, \boldsymbol{\xi}_k) > 0 \\ \text{for some } j \end{array}\right\}\right).$$

Algorithm 5.2 (Fuzzy Simulation for $U_2(\boldsymbol{x})$)
Step 1. Randomly generate θ_k from the credibility space $(\Theta, \mathcal{P}, \text{Cr})$, write $\nu_k = (2\text{Cr}\{\theta_k\}) \wedge 1$ and produce $\boldsymbol{\xi}_k = \boldsymbol{\xi}(\theta_k)$, $k = 1, 2, \cdots, N$, respectively. *Equivalently, randomly generate* $\boldsymbol{\xi}_k$ *and write* $\nu_k = \mu(\boldsymbol{\xi}_k)$ *for* $k = 1, 2, \cdots, N$, *where* μ *is the membership function of* $\boldsymbol{\xi}$.
Step 2. Return $U_2(\boldsymbol{x})$ via the estimation formula.

Fuzzy Simulation for $U_3(\boldsymbol{x})$

In order to compute the uncertain function $U_3(\boldsymbol{x})$, we randomly generate θ_k from the credibility space $(\Theta, \mathcal{P}, \mathrm{Cr})$, write $\nu_k = (2\mathrm{Cr}\{\theta_k\}) \wedge 1$ and produce $\boldsymbol{\xi}_k = \boldsymbol{\xi}(\theta_k)$, $k = 1, 2, \cdots, N$, respectively. *Equivalently, we randomly generate* $\boldsymbol{\xi}_k$ *and write* $\nu_k = \mu(\boldsymbol{\xi}_k)$ *for* $k = 1, 2, \cdots, N$, *where* μ *is the membership function of* $\boldsymbol{\xi}$. For any number r, we set

$$L(r) = \frac{1}{2}\left(\max_{1\le k\le N}\{\nu_k \mid f(\boldsymbol{x}, \boldsymbol{\xi}_k) \ge r\} + \min_{1\le k\le N}\{1 - \nu_k \mid f(\boldsymbol{x}, \boldsymbol{\xi}_k) < r\}\right).$$

It follows from monotonicity that we may employ bisection search to find the maximal value r such that $L(r) \ge \alpha$. This value is an estimation of $U_3(\boldsymbol{x})$. We summarize this process as follows.

Algorithm 5.3 (Fuzzy Simulation for $U_3(\boldsymbol{x})$)

Step 1. Randomly generate θ_k from the credibility space $(\Theta, \mathcal{P}, \mathrm{Cr})$, write $\nu_k = (2\mathrm{Cr}\{\theta_k\}) \wedge 1$ and produce $\boldsymbol{\xi}_k = \boldsymbol{\xi}(\theta_k)$, $k = 1, 2, \cdots, N$, respectively. *Equivalently, randomly generate* $\boldsymbol{\xi}_k$ *and write* $\nu_k = \mu(\boldsymbol{\xi}_k)$ *for* $k = 1, 2, \cdots, N$, *where* μ *is the membership function of* $\boldsymbol{\xi}$.

Step 2. Find the maximal value r such that $L(r) \ge \alpha$ holds.

Step 3. Return r.

Hybrid Intelligent Algorithm

Now we integrate fuzzy simulation, NN and GA to produce a hybrid intelligent algorithm for solving fuzzy programming models.

Algorithm 5.4 (Hybrid Intelligent Algorithm)

Step 1. Generate training input-output data for uncertain functions like

$$U_1 : \boldsymbol{x} \to E[f(\boldsymbol{x}, \boldsymbol{\xi})],$$
$$U_2 : \boldsymbol{x} \to \mathrm{Cr}\{g_j(\boldsymbol{x}, \boldsymbol{\xi}) \le 0, j = 1, 2, \cdots, p\},$$
$$U_3 : \boldsymbol{x} \to \max\left\{\overline{f} \mid \mathrm{Cr}\left\{f(\boldsymbol{x}, \boldsymbol{\xi}) \ge \overline{f}\right\} \ge \alpha\right\}$$

by the fuzzy simulation.

Step 2. Train a neural network to approximate the uncertain functions according to the generated training input-output data.

Step 3. Initialize *pop_size* chromosomes whose feasibility may be checked by the trained neural network.

Step 4. Update the chromosomes by crossover and mutation operations and the trained neural network may be employed to check the feasibility of offsprings.

Step 5. Calculate the objective values for all chromosomes by the trained neural network.
Step 6. Compute the fitness of each chromosome by rank-based evaluation function based on the objective values.
Step 7. Select the chromosomes by spinning the roulette wheel.
Step 8. Repeat the fourth to seventh steps a given number of cycles.
Step 9. Report the best chromosome as the optimal solution.

5.6 Numerical Experiments

In order to illustrate its effectiveness, a set of numerical examples has been done, and the results are successful. Here we give some numerical examples which are all performed on a personal computer with the following parameters: the population size is 30, the probability of crossover P_c is 0.3, the probability of mutation P_m is 0.2, and the parameter a in the rank-based evaluation function is 0.05.

Example 5.22. Consider first the following single-objective fuzzy EVM,

$$\begin{cases} \max E\left[\sqrt{|x_1+\xi_1|+|x_2+\xi_2|+|x_3+\xi_3|}\right] \\ \text{subject to:} \\ \quad x_1^2+x_2^2+x_3^2 \le 10 \end{cases}$$

where ξ_1, ξ_2 and ξ_3 are triangular fuzzy variables $(1,2,3)$, $(2,3,4)$, and $(3,4,5)$, respectively.

In order to solve this model, we first generate input-output data for the uncertain function

$$U : \boldsymbol{x} \to E\left[\sqrt{|x_1+\xi_1|+|x_2+\xi_2|+|x_3+\xi_3|}\right]$$

by fuzzy simulation. Then we train an NN (3 input neurons, 5 hidden neurons, 1 output neuron) to approximate the function $U(\boldsymbol{x})$. After that, the trained NN is embedded into a GA to produce a hybrid intelligent algorithm.

A run of the hybrid intelligent algorithm (6000 cycles in simulation, 2000 data in NN, 1000 generations in GA) shows that the optimal solution is

$$x_1^* = 1.8310, \quad x_2^* = 1.8417, \quad x_3^* = 1.8043$$

whose objective value is 3.80.

Example 5.23. Let us consider the following single-objective fuzzy CCP,

$$\begin{cases} \max \overline{f} \\ \text{subject to:} \\ \qquad \text{Cr}\left\{\sqrt{x_1+\xi_1}+\sqrt{x_2+\xi_2}+\sqrt{x_3+\xi_3}\ge \overline{f}\right\}\ge 0.9 \\ \qquad \text{Cr}\left\{\sqrt{(x_1+\xi_1)^2+(x_2+\xi_2)^2+(x_3+\xi_3)^2}\le 6\right\}\ge 0.8 \\ \qquad x_1,x_2,x_3\ge 0 \end{cases} \tag{5.42}$$

where ξ_1, ξ_2 and ξ_3 are assumed to triangular fuzzy variables $(0,1,2)$, $(1,2,3)$ and $(2,3,4)$, respectively.

In order to solve this model, we generate training input-output data for the uncertain function $U : \boldsymbol{x} \to (U_1(\boldsymbol{x}), U_2(\boldsymbol{x}))$, where

$$U_1(\boldsymbol{x}) = \max\left\{\overline{f} \mid \text{Cr}\left\{\sqrt{x_1+\xi_1}+\sqrt{x_2+\xi_2}+\sqrt{x_3+\xi_3}\ge \overline{f}\right\}\ge 0.9\right\},$$
$$U_2(\boldsymbol{x}) = \text{Cr}\left\{\sqrt{(x_1+\xi_1)^2+(x_2+\xi_2)^2+(x_3+\xi_3)^2}\le 6\right\}.$$

Then we train an NN (3 input neurons, 6 hidden neurons, 2 output neurons) to approximate the uncertain function U. Finally, we integrate the trained NN and GA to produce a hybrid intelligent algorithm.

A run of the hybrid intelligent algorithm (6000 cycles in simulation, 2000 training data in NN, 1500 generations in GA) shows that the optimal solution is

$$(x_1^*, x_2^*, x_3^*) = (1.9780, 0.6190, 0.0000)$$

with objective value $\overline{f}^* = 5.02$.

Example 5.24. We now consider the following CCGP model,

$$\begin{cases} \text{lexmin}\left\{d_1^- \vee 0, d_2^- \vee 0, d_3^- \vee 0\right\} \\ \text{subject to:} \\ \qquad \text{Cr}\left\{3-(x_1^2\xi_1+x_2\tau_1+x_3\eta_1^2)\le d_1^-\right\}\ge 0.90 \\ \qquad \text{Cr}\left\{4-(x_1\xi_2+x_2^2\tau_2^2+x_3\eta_2)\le d_2^-\right\}\ge 0.85 \\ \qquad \text{Cr}\left\{6-(x_1\xi_3^2+x_2\tau_3+x_3^2\eta_3)\le d_3^-\right\}\ge 0.80 \\ \qquad x_1+x_2+x_3=1 \\ \qquad x_1,x_2,x_3\ge 0 \end{cases} \tag{5.43}$$

where ξ_1, ξ_2, ξ_3 are fuzzy variables with membership functions $\exp[-|x-1|]$, $\exp[-|x-2|]$, $\exp[-|x-3|]$, τ_1, τ_2, τ_3 are triangular fuzzy variables $(1,2,3)$, $(2,3,4)$, $(3,4,5)$, η_1, η_2, η_3 are trapezoidal fuzzy variables $(2,3,4,5)$, $(3,4,5,6)$, $(4,5,6,7)$, respectively.

In order to solve this problem, we employ fuzzy simulation to generate input-output data for the uncertain function $U : \boldsymbol{x} \to (U_1(\boldsymbol{x}), U_2(\boldsymbol{x}), U_3(\boldsymbol{x}))$, where

$$
\begin{aligned}
U_1(\boldsymbol{x}) &= \max\left\{d \mid \mathrm{Cr}\left\{x_1^2\xi_1 + x_2\tau_1 + x_3\eta_1^2 \ge d\right\} \ge 0.90\right\},\\
U_2(\boldsymbol{x}) &= \max\left\{d \mid \mathrm{Cr}\left\{x_1\xi_2 + x_2^2\tau_2^2 + x_3\eta_2 \ge d\right\} \ge 0.85\right\},\\
U_3(\boldsymbol{x}) &= \max\left\{d \mid \mathrm{Cr}\left\{x_1\xi_3^2 + x_2\tau_3 + x_3^2\eta_3 \ge d\right\} \ge 0.80\right\}.
\end{aligned}
$$

Then we train an NN (3 input neurons, 8 hidden neurons, 3 output neurons) to approximate the uncertain function U. Note that

$$
d_1^- = [3 - U_1(\boldsymbol{x})] \vee 0, \quad d_2^- = [4 - U_2(\boldsymbol{x})] \vee 0, \quad d_3^- = [6 - U_3(\boldsymbol{x})] \vee 0.
$$

Finally, we integrate the trained NN and GA to produce a hybrid intelligent algorithm.

A run of the hybrid intelligent algorithm (5000 cycles in simulation, 3000 training data in NN, 3000 generations in GA) shows that the optimal solution is

$$
(x_1^*, x_2^*, x_3^*) = (0.2910, 0.5233, 0.1857)
$$

which can satisfy the first two goals, but the negative deviation of the third goal is 0.57.

Example 5.25. Let us now turn our attention to the following DCGP,

$$
\begin{cases}
\text{lexmin}\left\{d_1^- \vee 0, d_2^- \vee 0, d_3^- \vee 0\right\}\\
\text{subject to:}\\
\qquad 0.95 - \mathrm{Cr}\{x_1 + x_3^2 = 6\} = d_1^-\\
\qquad 0.90 - \mathrm{Cr}\{x_2 + x_5^2 = 5\} = d_2^-\\
\qquad 0.85 - \mathrm{Cr}\{x_4 + x_6^2 = 4\} = d_3^-\\
\qquad x_1 + x_2 \le \tilde{a}\\
\qquad x_3 + x_4 \le \tilde{b}\\
\qquad x_5 \le \tilde{c}\\
\qquad x_6 \le \tilde{d}\\
\qquad x_i \ge 0, \quad i = 1, 2, \cdots, 6
\end{cases}
$$

where $\tilde{a}, \tilde{b}, \tilde{c}$ are triangular fuzzy variables (3,4,5), (2,3,4), (0,1,2), respectively, and $\tilde{d}$ is a fuzzy variable with membership function $\mu_{\tilde{d}}(r) = 1/[1 + (r-1)^2]$.

In the first priority level, there is only one event denoted by $\mathcal{E}_1$ in the fuzzy environment, which should be fulfilled by $x_1 + x_3^2 = 6$. It is clear that the support $\mathcal{E}_1^* = \{x_1, x_3\}$ and the dependent support $\mathcal{E}_1^{**} = \{x_1, x_2, x_3, x_4\}$. It follows from the principle of uncertainty that the chance function $f_1(\boldsymbol{x})$ of the event $\mathcal{E}_1$ is

$$
f_1(\boldsymbol{x}) = \mathrm{Cr}\begin{Bmatrix}
x_1 + x_3^2 = 6\\
x_1 + x_2 \le \tilde{a}\\
x_3 + x_4 \le \tilde{b}\\
x_1, x_2, x_3, x_4 \ge 0
\end{Bmatrix}.
$$

At the second priority level, there is an event $\mathcal{E}_2$ which will be fulfilled by $x_2 + x_5^2 = 5$. The support $\mathcal{E}_2^* = \{x_2, x_5\}$ and the dependent support $\mathcal{E}_2^{**} = \{x_1, x_2, x_5\}$. It follows from the principle of uncertainty that the chance function $f_2(\boldsymbol{x})$ of the event $\mathcal{E}_2$ is

$$f_2(\boldsymbol{x}) = \text{Cr}\left\{\begin{array}{l} x_2 + x_5^2 = 5 \\ x_1 + x_2 \le \tilde{a} \\ x_5 \le \tilde{c} \\ x_1, x_2, x_5 \ge 0 \end{array}\right\}.$$

At the third priority level, there is an event $\mathcal{E}_3$ which will be fulfilled by $x_4 + x_6^2 = 4$. The support $\mathcal{E}_3^* = \{x_4, x_6\}$ and the dependent support $\mathcal{E}_3^{**} = \{x_3, x_4, x_6\}$. It follows from the principle of uncertainty that the chance function $f_3(\boldsymbol{x})$ of the event $\mathcal{E}_3$ is

$$f_3(\boldsymbol{x}) = \text{Cr}\left\{\begin{array}{l} x_4 + x_6^2 = 4 \\ x_3 + x_4 \le \tilde{b} \\ x_6 \le \tilde{d} \\ x_3, x_4, x_6 \ge 0 \end{array}\right\}.$$

We encode a solution by a chromosome $V = (v_1, v_2, v_3)$, and decode it into a feasible solution in the following way,

$$\begin{array}{ll} x_1 = v_1, \ x_2 = v_2, & x_3 = \sqrt{6 - v_1} \\ x_4 = v_3, \ x_5 = \sqrt{5 - v_2}, & x_6 = \sqrt{4 - v_3} \end{array}$$

which ensures that $x_1 + x_3^2 = 6$, $x_2 + x_5^2 = 5$ and $x_4 + x_6^2 = 4$.

At first, we employ fuzzy simulation to generate input-output data for the chance function

$$U : (v_1, v_2, v_3) \to (f_1(\boldsymbol{x}), f_2(\boldsymbol{x}), f_3(\boldsymbol{x})).$$

Then we train an NN (3 input neurons, 8 hidden neurons, 3 output neurons) to approximate it. Finally, we embed the trained NN into a GA to produce a hybrid intelligent algorithm.

A run of the hybrid intelligent algorithm (5000 cycles in simulation, 2000 data in NN, 1000 generations in GA) shows that the optimal solution is

$$\boldsymbol{x}^* = (0.2097, 3.8263, 2.4063, 0.6407, 1.0833, 1.8328)$$

which can satisfy the first and second goals, but the third objective is 0.25.

Chapter 6
Hybrid Programming

In many cases, fuzziness and randomness simultaneously appear in a system. In order to describe this phenomena, a fuzzy random variable was introduced by Kwakernaak [142] as a random element taking "fuzzy variable" values. By fuzzy random programming we mean the optimization theory in fuzzy random environments. Liu and Liu [198] presented a spectrum of fuzzy random expected value model (EVM), Liu [179] initialized a general framework of fuzzy random chance-constrained programming (CCP), and Liu [180] introduced the concepts of uncertain environment and chance function for fuzzy random decision problems, and constructed a theoretical framework of fuzzy random dependent-chance programming (DCP).

A random fuzzy variable was proposed by Liu [181] as a fuzzy element taking "random variable" values. By random fuzzy programming we mean the optimization theory in random fuzzy environments. Liu and Liu [200] introduced a spectrum of random fuzzy EVM, Liu [181] proposed the random fuzzy CCP, and Liu [185] presented a spectrum of random fuzzy DCP in which the underlying philosophy is based on selecting the decision with maximum chance to meet the event.

More generally, a hybrid variable was introduced by Liu [187] as a measurable function from a chance space to the set of real numbers. Fuzzy random variable and random fuzzy variable are instances of hybrid variable. In order to measure hybrid events, a concept of chance measure was introduced by Li and Liu [161]. This chapter will assume the hybrid environment and introduce a spectrum of hybrid programming. In order to solve general hybrid programming, we will integrate hybrid simulation, neural network (NN) and genetic algorithm (GA) to produce a hybrid intelligent algorithm, and illustrate its effectiveness via some numerical examples.

6.1 Hybrid Variables

Let us start this section with the concept of chance space. Essentially, a chance space is the product of credibility space and probability space.

B. Liu: Theory and Practice of Uncertain Programming, STUDFUZZ 239, pp. 83–110.
springerlink.com

Definition 6.1. *(Liu [187]) Suppose that* $(\Theta, \mathcal{P}, \mathrm{Cr})$ *is a credibility space and* $(\Omega, \mathcal{A}, \Pr)$ *is a probability space. The product* $(\Theta, \mathcal{P}, \mathrm{Cr}) \times (\Omega, \mathcal{A}, \Pr)$ *is called a chance space.*

The universal set $\Theta \times \Omega$ is clearly the set of all ordered pairs of the form (θ, ω), where $\theta \in \Theta$ and $\omega \in \Omega$. What is the product σ-algebra $\mathcal{P} \times \mathcal{A}$? What is the product measure $\mathrm{Cr} \times \Pr$? Let us discuss these two basic problems.

What Is the Product σ-Algebra $\mathcal{P} \times \mathcal{A}$?

Definition 6.2. *(Liu [189]) Let* $(\Theta, \mathcal{P}, \mathrm{Cr}) \times (\Omega, \mathcal{A}, \Pr)$ *be a chance space. A subset* $\Lambda \subset \Theta \times \Omega$ *is called an event if*

$$\Lambda(\theta) = \big\{\omega \in \Omega \bigm| (\theta, \omega) \in \Lambda\big\} \in \mathcal{A} \tag{6.1}$$

for each $\theta \in \Theta$.

Example 6.1. Empty set $\emptyset$ and universal set $\Theta \times \Omega$ are clearly events.

Example 6.2. Let $X \in \mathcal{P}$ and $Y \in \mathcal{A}$. Then $X \times Y$ is a subset of $\Theta \times \Omega$. Since the set

$$(X \times Y)(\theta) = \begin{cases} Y, \text{ if } \theta \in X \\ \emptyset, \text{ if } \theta \in X^c \end{cases}$$

is in the σ-algebra $\mathcal{A}$ for each $\theta \in \Theta$, the rectangle $X \times Y$ is an event.

Theorem 6.1. *(Liu [189]) Let* $(\Theta, \mathcal{P}, \mathrm{Cr}) \times (\Omega, \mathcal{A}, \Pr)$ *be a chance space. The class of all events is a* σ*-algebra over* $\Theta \times \Omega$*, and denoted by* $\mathcal{P} \times \mathcal{A}$.

Proof: At first, it is obvious that $\Theta \times \Omega \in \mathcal{P} \times \mathcal{A}$. For any event Λ, we always have

$$\Lambda(\theta) \in \mathcal{A}, \quad \forall \theta \in \Theta.$$

Thus for each $\theta \in \Theta$, the set

$$\Lambda^c(\theta) = \big\{\omega \in \Omega \bigm| (\theta, \omega) \in \Lambda^c\big\} = (\Lambda(\theta))^c \in \mathcal{A}$$

which implies that $\Lambda^c \in \mathcal{P} \times \mathcal{A}$. Finally, let $\Lambda_1, \Lambda_2, \cdots$ be events. Then for each $\theta \in \Theta$, we have

$$\left(\bigcup_{i=1}^{\infty} \Lambda_i\right)(\theta) = \left\{\omega \in \Omega \mid (\theta, \omega) \in \bigcup_{i=1}^{\infty} \Lambda_i\right\} = \bigcup_{i=1}^{\infty} \big\{\omega \in \Omega \bigm| (\theta, \omega) \in \Lambda_i\big\} \in \mathcal{A}.$$

That is, the countable union $\cup_i \Lambda_i \in \mathcal{P} \times \mathcal{A}$. Hence $\mathcal{P} \times \mathcal{A}$ is a σ-algebra.

What Is the Product Measure Cr × Pr?

Product probability is a probability measure, and product credibility is a credibility measure. What is the product measure Cr × Pr? We will call it *chance measure* and define it as follows.

Definition 6.3. *(Li and Liu [161]) Let* $(\Theta, \mathcal{P}, \text{Cr}) \times (\Omega, \mathcal{A}, \text{Pr})$ *be a chance space. Then a chance measure of an event* Λ *is defined as*

$$\text{Ch}\{\Lambda\} = \begin{cases} \sup\limits_{\theta\in\Theta}(\text{Cr}\{\theta\} \wedge \Pr\{\Lambda(\theta)\}), \\ \qquad \text{if } \sup\limits_{\theta\in\Theta}(\text{Cr}\{\theta\} \wedge \Pr\{\Lambda(\theta)\}) < 0.5 \\ 1 - \sup\limits_{\theta\in\Theta}(\text{Cr}\{\theta\} \wedge \Pr\{\Lambda^c(\theta)\}), \\ \qquad \text{if } \sup\limits_{\theta\in\Theta}(\text{Cr}\{\theta\} \wedge \Pr\{\Lambda(\theta)\}) \geq 0.5. \end{cases} \tag{6.2}$$

Theorem 6.2. *Let* $(\Theta, \mathcal{P}, \text{Cr}) \times (\Omega, \mathcal{A}, \text{Pr})$ *be a chance space and* Ch *a chance measure. Then we have*

$$\text{Ch}\{\emptyset\} = 0, \tag{6.3}$$

$$\text{Ch}\{\Theta \times \Omega\} = 1, \tag{6.4}$$

$$0 \leq \text{Ch}\{\Lambda\} \leq 1 \tag{6.5}$$

for any event Λ.

Proof: It follows from the definition immediately.

Theorem 6.3. *Let* $(\Theta, \mathcal{P}, \text{Cr}) \times (\Omega, \mathcal{A}, \text{Pr})$ *be a chance space and* Ch *a chance measure. Then for any event* Λ*, we have*

$$\sup_{\theta\in\Theta}(\text{Cr}\{\theta\} \wedge \Pr\{\Lambda(\theta)\}) \vee \sup_{\theta\in\Theta}(\text{Cr}\{\theta\} \wedge \Pr\{\Lambda^c(\theta)\}) \geq 0.5, \tag{6.6}$$

$$\sup_{\theta\in\Theta}(\text{Cr}\{\theta\} \wedge \Pr\{\Lambda(\theta)\}) + \sup_{\theta\in\Theta}(\text{Cr}\{\theta\} \wedge \Pr\{\Lambda^c(\theta)\}) \leq 1, \tag{6.7}$$

$$\sup_{\theta\in\Theta}(\text{Cr}\{\theta\} \wedge \Pr\{\Lambda(\theta)\}) \leq \text{Ch}\{\Lambda\} \leq 1 - \sup_{\theta\in\Theta}(\text{Cr}\{\theta\} \wedge \Pr\{\Lambda^c(\theta)\}). \tag{6.8}$$

Proof: It follows from the basic properties of probability and credibility that

$$\begin{aligned} &\sup_{\theta\in\Theta}(\text{Cr}\{\theta\} \wedge \Pr\{\Lambda(\theta)\}) \vee \sup_{\theta\in\Theta}(\text{Cr}\{\theta\} \wedge \Pr\{\Lambda^c(\theta)\}) \\ &\geq \sup_{\theta\in\Theta}(\text{Cr}\{\theta\} \wedge (\Pr\{\Lambda(\theta)\} \vee \Pr\{\Lambda^c(\theta)\})) \\ &\geq \sup_{\theta\in\Theta}\text{Cr}\{\theta\} \wedge 0.5 = 0.5 \end{aligned}$$

and

$$\begin{aligned}
&\sup_{\theta\in\Theta}(\mathrm{Cr}\{\theta\}\wedge\Pr\{\Lambda(\theta)\})+\sup_{\theta\in\Theta}(\mathrm{Cr}\{\theta\}\wedge\Pr\{\Lambda^c(\theta)\})\\
=&\sup_{\theta_1,\theta_2\in\Theta}(\mathrm{Cr}\{\theta_1\}\wedge\Pr\{\Lambda(\theta_1)\}+\mathrm{Cr}\{\theta_2\}\wedge\Pr\{\Lambda^c(\theta_2)\})\\
\le&\sup_{\theta_1\ne\theta_2}(\mathrm{Cr}\{\theta_1\}+\mathrm{Cr}\{\theta_2\})\vee\sup_{\theta\in\Theta}(\Pr\{\Lambda(\theta)\}+\Pr\{\Lambda^c(\theta)\})\\
\le&\,1\vee 1=1.
\end{aligned}$$

The inequalities (6.8) follows immediately from the above inequalities and the definition of chance measure.

Theorem 6.4. *(Li and Liu [161]) The chance measure is increasing. That is,*

$$\mathrm{Ch}\{\Lambda_1\}\le\mathrm{Ch}\{\Lambda_2\}\tag{6.9}$$

for any events Λ_1 and Λ_2 with $\Lambda_1\subset\Lambda_2$.

Proof: Since $\Lambda_1(\theta)\subset\Lambda_2(\theta)$ and $\Lambda_2^c(\theta)\subset\Lambda_1^c(\theta)$ for each $\theta\in\Theta$, we have

$$\sup_{\theta\in\Theta}(\mathrm{Cr}\{\theta\}\wedge\Pr\{\Lambda_1(\theta)\})\le\sup_{\theta\in\Theta}(\mathrm{Cr}\{\theta\}\wedge\Pr\{\Lambda_2(\theta)\}),$$

$$\sup_{\theta\in\Theta}(\mathrm{Cr}\{\theta\}\wedge\Pr\{\Lambda_2^c(\theta)\})\le\sup_{\theta\in\Theta}(\mathrm{Cr}\{\theta\}\wedge\Pr\{\Lambda_1^c(\theta)\}).$$

The argument breaks down into three cases.

Case 1: $\sup_{\theta\in\Theta}(\mathrm{Cr}\{\theta\}\wedge\Pr\{\Lambda_2(\theta)\})<0.5$. For this case, we have

$$\sup_{\theta\in\Theta}(\mathrm{Cr}\{\theta\}\wedge\Pr\{\Lambda_1(\theta)\})<0.5,$$

$$\mathrm{Ch}\{\Lambda_2\}=\sup_{\theta\in\Theta}(\mathrm{Cr}\{\theta\}\wedge\Pr\{\Lambda_2(\theta)\})\ge\sup_{\theta\in\Theta}(\mathrm{Cr}\{\theta\}\wedge\Pr\{\Lambda_1(\theta)\}=\mathrm{Ch}\{\Lambda_1\}.$$

Case 2: $\sup_{\theta\in\Theta}(\mathrm{Cr}\{\theta\}\wedge\Pr\{\Lambda_2(\theta)\})\ge 0.5$ and $\sup_{\theta\in\Theta}(\mathrm{Cr}\{\theta\}\wedge\Pr\{\Lambda_1(\theta)\})<0.5$. It follows from Theorem 6.3 that

$$\mathrm{Ch}\{\Lambda_2\}\ge\sup_{\theta\in\Theta}(\mathrm{Cr}\{\theta\}\wedge\Pr\{\Lambda_2(\theta)\})\ge 0.5>\mathrm{Ch}\{\Lambda_1\}.$$

Case 3: $\sup_{\theta\in\Theta}(\mathrm{Cr}\{\theta\}\wedge\Pr\{\Lambda_2(\theta)\})\ge 0.5$ and $\sup_{\theta\in\Theta}(\mathrm{Cr}\{\theta\}\wedge\Pr\{\Lambda_1(\theta)\})\ge 0.5$. For this case, we have

$$\mathrm{Ch}\{\Lambda_2\}=1-\sup_{\theta\in\Theta}(\mathrm{Cr}\{\theta\}\wedge\Pr\{\Lambda_2^c(\theta)\})\ge 1-\sup_{\theta\in\Theta}(\mathrm{Cr}\{\theta\}\wedge\Pr\{\Lambda_1^c(\theta)\})=\mathrm{Ch}\{\Lambda_1\}.$$

Thus Ch is an increasing measure.

Theorem 6.5. *(Li and Liu [161]) The chance measure is self-dual. That is,*

$$\mathrm{Ch}\{\Lambda\} + \mathrm{Ch}\{\Lambda^c\} = 1 \tag{6.10}$$

for any event Λ.

Proof: For any event Λ, please note that

$$\mathrm{Ch}\{\Lambda^c\} = \begin{cases} \sup\limits_{\theta\in\Theta}(\mathrm{Cr}\{\theta\}\wedge\mathrm{Pr}\{\Lambda^c(\theta)\}), & \text{if } \sup\limits_{\theta\in\Theta}(\mathrm{Cr}\{\theta\}\wedge\mathrm{Pr}\{\Lambda^c(\theta)\}) < 0.5 \\ 1-\sup\limits_{\theta\in\Theta}(\mathrm{Cr}\{\theta\}\wedge\mathrm{Pr}\{\Lambda(\theta)\}), & \text{if } \sup\limits_{\theta\in\Theta}(\mathrm{Cr}\{\theta\}\wedge\mathrm{Pr}\{\Lambda^c(\theta)\}) \ge 0.5. \end{cases}$$

The argument breaks down into three cases.

Case 1: $\sup\limits_{\theta\in\Theta}(\mathrm{Cr}\{\theta\}\wedge\mathrm{Pr}\{\Lambda(\theta)\}) < 0.5$. For this case, we have

$$\sup_{\theta\in\Theta}(\mathrm{Cr}\{\theta\}\wedge\mathrm{Pr}\{\Lambda^c(\theta)\}) \ge 0.5,$$

$$\mathrm{Ch}\{\Lambda\}+\mathrm{Ch}\{\Lambda^c\} = \sup_{\theta\in\Theta}(\mathrm{Cr}\{\theta\}\wedge\mathrm{Pr}\{\Lambda(\theta)\}) + 1 - \sup_{\theta\in\Theta}(\mathrm{Cr}\{\theta\}\wedge\mathrm{Pr}\{\Lambda(\theta)\}) = 1.$$

Case 2: $\sup\limits_{\theta\in\Theta}(\mathrm{Cr}\{\theta\}\wedge\mathrm{Pr}\{\Lambda(\theta)\}) \ge 0.5$ and $\sup\limits_{\theta\in\Theta}(\mathrm{Cr}\{\theta\}\wedge\mathrm{Pr}\{\Lambda^c(\theta)\}) < 0.5$. For this case, we have

$$\mathrm{Ch}\{\Lambda\}+\mathrm{Ch}\{\Lambda^c\} = 1 - \sup_{\theta\in\Theta}(\mathrm{Cr}\{\theta\}\wedge\mathrm{Pr}\{\Lambda^c(\theta)\}) + \sup_{\theta\in\Theta}(\mathrm{Cr}\{\theta\}\wedge\mathrm{Pr}\{\Lambda^c(\theta)\}) = 1.$$

Case 3: $\sup\limits_{\theta\in\Theta}(\mathrm{Cr}\{\theta\}\wedge\mathrm{Pr}\{\Lambda(\theta)\}) \ge 0.5$ and $\sup\limits_{\theta\in\Theta}(\mathrm{Cr}\{\theta\}\wedge\mathrm{Pr}\{\Lambda^c(\theta)\}) \ge 0.5$. For this case, it follows from Theorem 6.3 that

$$\sup_{\theta\in\Theta}(\mathrm{Cr}\{\theta\}\wedge\mathrm{Pr}\{\Lambda(\theta)\}) = \sup_{\theta\in\Theta}(\mathrm{Cr}\{\theta\}\wedge\mathrm{Pr}\{\Lambda^c(\theta)\}) = 0.5.$$

Hence $\mathrm{Ch}\{\Lambda\}+\mathrm{Ch}\{\Lambda^c\} = 0.5+0.5 = 1$. The theorem is proved.

Theorem 6.6. *(Li and Liu [161]) For any event $X\times Y$, we have*

$$\mathrm{Ch}\{X\times Y\} = \mathrm{Cr}\{X\}\wedge\mathrm{Pr}\{Y\}. \tag{6.11}$$

Proof: The argument breaks down into three cases.

Case 1: $\mathrm{Cr}\{X\} < 0.5$. For this case, we have

$$\sup_{\theta\in X}\mathrm{Cr}\{\theta\}\wedge\mathrm{Pr}\{Y\} = \mathrm{Cr}\{X\}\wedge\mathrm{Cr}\{Y\} < 0.5,$$

$$\mathrm{Ch}\{X\times Y\} = \sup_{\theta\in X}\mathrm{Cr}\{\theta\}\wedge\mathrm{Pr}\{Y\} = \mathrm{Cr}\{X\}\wedge\mathrm{Pr}\{Y\}.$$

Case 2: $\mathrm{Cr}\{X\} \ge 0.5$ and $\mathrm{Pr}\{Y\} < 0.5$. Then we have

$$\sup_{\theta\in X}\mathrm{Cr}\{\theta\} \ge 0.5,$$

$$\sup_{\theta\in X} \mathrm{Cr}\{\theta\} \wedge \Pr\{Y\} = \Pr\{Y\} < 0.5,$$

$$\mathrm{Ch}\{X \times Y\} = \sup_{\theta\in X} \mathrm{Cr}\{\theta\} \wedge \Pr\{Y\} = \Pr\{Y\} = \mathrm{Cr}\{X\} \wedge \Pr\{Y\}.$$

Case 3: $\mathrm{Cr}\{X\} \geq 0.5$ and $\Pr\{Y\} \geq 0.5$. Then we have

$$\sup_{\theta\in\Theta} (\mathrm{Cr}\{\theta\} \wedge \Pr\{(X \times Y)(\theta)\}) \geq \sup_{\theta\in X} \mathrm{Cr}\{\theta\} \wedge \Pr\{Y\} \geq 0.5,$$

$$\mathrm{Ch}\{X \times Y\} = 1 - \sup_{\theta\in\Theta} (\mathrm{Cr}\{\theta\} \wedge \Pr\{(X \times Y)^c(\theta)\}) = \mathrm{Cr}\{X\} \wedge \Pr\{Y\}.$$

The theorem is proved.

Example 6.3. It follows from Theorem 6.6 that for any events $X \times \Omega$ and $\Theta \times Y$, we have

$$\mathrm{Ch}\{X \times \Omega\} = \mathrm{Cr}\{X\}, \quad \mathrm{Ch}\{\Theta \times Y\} = \Pr\{Y\}. \tag{6.12}$$

Theorem 6.7. *(Li and Liu [161], Chance Subadditivity Theorem) The chance measure is subadditive. That is,*

$$\mathrm{Ch}\{\Lambda_1 \cup \Lambda_2\} \leq \mathrm{Ch}\{\Lambda_1\} + \mathrm{Ch}\{\Lambda_2\} \tag{6.13}$$

for any events Λ_1 and Λ_2. In fact, chance measure is not only finitely subadditive but also countably subadditive.

Proof: The proof breaks down into three cases.

Case 1: $\mathrm{Ch}\{\Lambda_1 \cup \Lambda_2\} < 0.5$. Then $\mathrm{Ch}\{\Lambda_1\} < 0.5$, $\mathrm{Ch}\{\Lambda_2\} < 0.5$ and

$$\begin{aligned}
\mathrm{Ch}\{\Lambda_1 \cup \Lambda_2\} &= \sup_{\theta\in\Theta}(\mathrm{Cr}\{\theta\} \wedge \Pr\{(\Lambda_1 \cup \Lambda_2)(\theta)\}) \\
&\leq \sup_{\theta\in\Theta}(\mathrm{Cr}\{\theta\} \wedge (\Pr\{\Lambda_1(\theta)\} + \Pr\{\Lambda_2(\theta)\})) \\
&\leq \sup_{\theta\in\Theta}(\mathrm{Cr}\{\theta\} \wedge \Pr\{\Lambda_1(\theta)\} + \mathrm{Cr}\{\theta\} \wedge \Pr\{\Lambda_2(\theta)\}) \\
&\leq \sup_{\theta\in\Theta}(\mathrm{Cr}\{\theta\} \wedge \Pr\{\Lambda_1(\theta)\}) + \sup_{\theta\in\Theta}(\mathrm{Cr}\{\theta\} \wedge \Pr\{\Lambda_2(\theta)\}) \\
&= \mathrm{Ch}\{\Lambda_1\} + \mathrm{Ch}\{\Lambda_2\}.
\end{aligned}$$

Case 2: $\mathrm{Ch}\{\Lambda_1 \cup \Lambda_2\} \geq 0.5$ and $\mathrm{Ch}\{\Lambda_1\} \vee \mathrm{Ch}\{\Lambda_2\} < 0.5$. We first have

$$\sup_{\theta\in\Theta}(\mathrm{Cr}\{\theta\} \wedge \Pr\{(\Lambda_1 \cup \Lambda_2)(\theta)\}) \geq 0.5.$$

For any sufficiently small number $\varepsilon > 0$, there exists a point θ such that

$$\mathrm{Cr}\{\theta\} \wedge \Pr\{(\Lambda_1 \cup \Lambda_2)(\theta)\} > 0.5 - \varepsilon > \mathrm{Ch}\{\Lambda_1\} \vee \mathrm{Ch}\{\Lambda_2\},$$

$$\mathrm{Cr}\{\theta\} > 0.5 - \varepsilon > \Pr\{\Lambda_1(\theta)\},$$

$$\mathrm{Cr}\{\theta\} > 0.5 - \varepsilon > \Pr\{\Lambda_2(\theta)\}.$$

Thus we have

$$\begin{aligned}&\text{Cr}\{\theta\} \wedge \Pr\{(\Lambda_1 \cup \Lambda_2)^c(\theta)\} + \text{Cr}\{\theta\} \wedge \Pr\{\Lambda_1(\theta)\} + \text{Cr}\{\theta\} \wedge \Pr\{\Lambda_2(\theta)\}\\ =\ &\text{Cr}\{\theta\} \wedge \Pr\{(\Lambda_1 \cup \Lambda_2)^c(\theta)\} + \Pr\{\Lambda_1(\theta)\} + \Pr\{\Lambda_2(\theta)\}\\ \ge\ &\text{Cr}\{\theta\} \wedge \Pr\{(\Lambda_1 \cup \Lambda_2)^c(\theta)\} + \Pr\{(\Lambda_1 \cup \Lambda_2)(\theta)\} \ge 1 - 2\varepsilon\end{aligned}$$

because if $\text{Cr}\{\theta\} \ge \Pr\{(\Lambda_1 \cup \Lambda_2)^c(\theta)\}$, then

$$\begin{aligned}&\text{Cr}\{\theta\} \wedge \Pr\{(\Lambda_1 \cup \Lambda_2)^c(\theta)\} + \Pr\{(\Lambda_1 \cup \Lambda_2)(\theta)\}\\ =\ &\Pr\{(\Lambda_1 \cup \Lambda_2)^c(\theta)\} + \Pr\{(\Lambda_1 \cup \Lambda_2)(\theta)\}\\ =\ &1 \ge 1 - 2\varepsilon\end{aligned}$$

and if $\text{Cr}\{\theta\} < \Pr\{(\Lambda_1 \cup \Lambda_2)^c(\theta)\}$, then

$$\begin{aligned}&\text{Cr}\{\theta\} \wedge \Pr\{(\Lambda_1 \cup \Lambda_2)^c(\theta)\} + \Pr\{(\Lambda_1 \cup \Lambda_2)(\theta)\}\\ =\ &\text{Cr}\{\theta\} + \Pr\{(\Lambda_1 \cup \Lambda_2)(\theta)\}\\ \ge\ &(0.5 - \varepsilon) + (0.5 - \varepsilon) = 1 - 2\varepsilon.\end{aligned}$$

Taking supremum on both sides and letting $\varepsilon \to 0$, we obtain

$$\begin{aligned}\text{Ch}\{\Lambda_1 \cup \Lambda_2\} &= 1 - \sup_{\theta\in\Theta}(\text{Cr}\{\theta\} \wedge \Pr\{(\Lambda_1 \cup \Lambda_2)^c(\theta)\})\\ &\le \sup_{\theta\in\Theta}(\text{Cr}\{\theta\} \wedge \Pr\{\Lambda_1(\theta)\}) + \sup_{\theta\in\Theta}(\text{Cr}\{\theta\} \wedge \Pr\{\Lambda_2(\theta)\})\\ &= \text{Ch}\{\Lambda_1\} + \text{Ch}\{\Lambda_2\}.\end{aligned}$$

Case 3: $\text{Ch}\{\Lambda_1 \cup \Lambda_2\} \ge 0.5$ and $\text{Ch}\{\Lambda_1\} \vee \text{Ch}\{\Lambda_2\} \ge 0.5$. Without loss of generality, suppose $\text{Ch}\{\Lambda_1\} \ge 0.5$. For each θ, we first have

$$\begin{aligned}\text{Cr}\{\theta\} \wedge \Pr\{\Lambda_1^c(\theta)\} &= \text{Cr}\{\theta\} \wedge \Pr\{(\Lambda_1^c(\theta) \cap \Lambda_2^c(\theta)) \cup (\Lambda_1^c(\theta) \cap \Lambda_2(\theta))\}\\ &\le \text{Cr}\{\theta\} \wedge (\Pr\{(\Lambda_1 \cup \Lambda_2)^c(\theta)\} + \Pr\{\Lambda_2(\theta)\})\\ &\le \text{Cr}\{\theta\} \wedge \Pr\{(\Lambda_1 \cup \Lambda_2)^c(\theta)\} + \text{Cr}\{\theta\} \wedge \Pr\{\Lambda_2(\theta)\},\end{aligned}$$

i.e., $\text{Cr}\{\theta\} \wedge \Pr\{(\Lambda_1 \cup \Lambda_2)^c(\theta)\} \ge \text{Cr}\{\theta\} \wedge \Pr\{\Lambda_1^c(\theta)\} - \text{Cr}\{\theta\} \wedge \Pr\{\Lambda_2(\theta)\}$. It follows from Theorem 6.3 that

$$\begin{aligned}\text{Ch}\{\Lambda_1 \cup \Lambda_2\} &= 1 - \sup_{\theta\in\Theta}(\text{Cr}\{\theta\} \wedge \Pr\{(\Lambda_1 \cup \Lambda_2)^c(\theta)\})\\ &\le 1 - \sup_{\theta\in\Theta}(\text{Cr}\{\theta\} \wedge \Pr\{\Lambda_1^c(\theta)\}) + \sup_{\theta\in\Theta}(\text{Cr}\{\theta\} \wedge \Pr\{\Lambda_2(\theta)\})\\ &\le \text{Ch}\{\Lambda_1\} + \text{Ch}\{\Lambda_2\}.\end{aligned}$$

The theorem is proved.

Hybrid Variables

Recall that a random variable is a measurable function from a probability space to the set of real numbers, and a fuzzy variable is a function from a credibility space to the set of real numbers. In order to describe a quantity with both fuzziness and randomness, we introduce the concept of hybrid variable as follows.

Definition 6.4. *(Liu [187]) A hybrid variable is a measurable function from a chance space* $(\Theta, \mathcal{P}, \mathrm{Cr}) \times (\Omega, \mathcal{A}, \Pr)$ *to the set of real numbers, i.e., for any Borel set* B *of real numbers, the set*

$$\{\xi \in B\} = \{(\theta, \omega) \in \Theta \times \Omega \mid \xi(\theta, \omega) \in B\} \tag{6.14}$$

is an event.

Remark 6.1. A hybrid variable degenerates to a fuzzy variable if the value of $\xi(\theta, \omega)$ does not vary with ω. For example,

$$\xi(\theta, \omega) = \theta, \quad \xi(\theta, \omega) = \theta^2 + 1, \quad \xi(\theta, \omega) = \sin\theta.$$

Remark 6.2. A hybrid variable degenerates to a random variable if the value of $\xi(\theta, \omega)$ does not vary with θ. For example,

$$\xi(\theta, \omega) = \omega, \quad \xi(\theta, \omega) = \omega^2 + 1, \quad \xi(\theta, \omega) = \sin\omega.$$

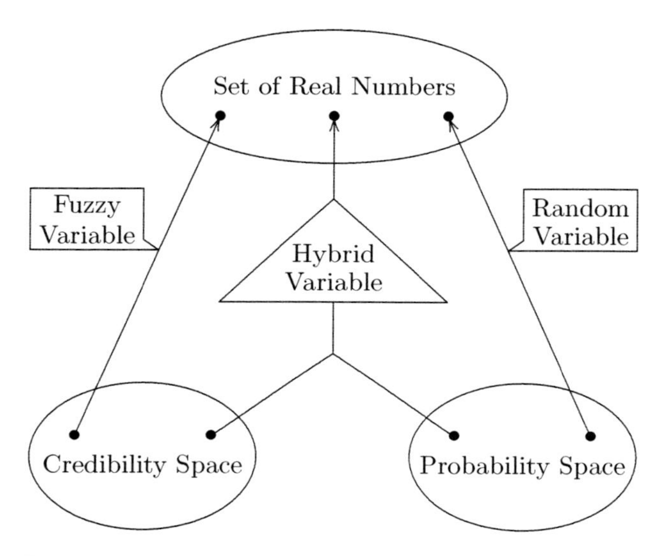

Fig. 6.1 Graphical Representation of Hybrid Variable

Remark 6.3. A hybrid variable $\xi(\theta, \omega)$ may also be regarded as a function from a credibility space $(\Theta, \mathcal{P}, \text{Cr})$ to the set $\{\xi(\theta, \cdot)|\theta \in \Theta\}$ of random variables. Thus ξ is a *random fuzzy variable* defined by Liu [181].

Remark 6.4. A hybrid variable $\xi(\theta, \omega)$ may be regarded as a function from a probability space $(\Omega, \mathcal{A}, \text{Pr})$ to the set $\{\xi(\cdot, \omega)|\omega \in \Omega\}$ of fuzzy variables. If $\text{Cr}\{\xi(\cdot, \omega) \in B\}$ is a measurable function of ω for any Borel set B of real numbers, then ξ is a *fuzzy random variable* in the sense of Liu and Liu [199].

Model I

If $\tilde{a}$ is a fuzzy variable and η is a random variable, then the sum $\xi = \tilde{a} + \eta$ is a hybrid variable. The product $\xi = \tilde{a} \cdot \eta$ is also a hybrid variable. Generally speaking, if $f : \Re^2 \to \Re$ is a measurable function, then

$$\xi = f(\tilde{a}, \eta) \tag{6.15}$$

is a hybrid variable. Suppose that $\tilde{a}$ has a membership function μ, and η has a probability density function ϕ. Then for any Borel set B of real numbers, we have

$$\text{Ch}\{f(\tilde{a}, \eta) \in B\} = \begin{cases} \sup\limits_x \left(\dfrac{\mu(x)}{2} \wedge \displaystyle\int_{f(x,y) \in B} \phi(y) \text{d}y \right), \\ \qquad \text{if } \sup\limits_x \left(\dfrac{\mu(x)}{2} \wedge \displaystyle\int_{f(x,y) \in B} \phi(y) \text{d}y \right) < 0.5 \\ 1 - \sup\limits_x \left(\dfrac{\mu(x)}{2} \wedge \displaystyle\int_{f(x,y) \in B^c} \phi(y) \text{d}y \right), \\ \qquad \text{if } \sup\limits_x \left(\dfrac{\mu(x)}{2} \wedge \displaystyle\int_{f(x,y) \in B} \phi(y) \text{d}y \right) \geq 0.5. \end{cases}$$

More generally, let $\tilde{a}_1, \tilde{a}_2, \cdots, \tilde{a}_m$ be fuzzy variables, and let $\eta_1, \eta_2, \cdots, \eta_n$ be random variables. If $f : \Re^{m+n} \to \Re$ is a measurable function, then

$$\xi = f(\tilde{a}_1, \tilde{a}_2, \cdots, \tilde{a}_m; \eta_1, \eta_2, \cdots, \eta_n) \tag{6.16}$$

is a hybrid variable. The chance $\text{Ch}\{f(\tilde{a}_1, \tilde{a}_2, \cdots, \tilde{a}_m; \eta_1, \eta_2, \cdots, \eta_n) \in B\}$ may be calculated in a similar way provided that μ is the joint membership function and ϕ is the joint probability density function.

Model II

Let $\tilde{a}_1, \tilde{a}_2, \cdots, \tilde{a}_m$ be fuzzy variables, and let $p_1, p_2, \cdots, p_m$ be nonnegative numbers with $p_1 + p_2 + \cdots + p_m = 1$. Then

$$\xi = \begin{cases} \tilde{a}_1 \text{ with probability } p_1 \\ \tilde{a}_2 \text{ with probability } p_2 \\ \cdots \\ \tilde{a}_m \text{ with probability } p_m \end{cases} \tag{6.17}$$

is clearly a hybrid variable. If $\tilde{a}_1, \tilde{a}_2, \cdots, \tilde{a}_m$ have membership functions $\mu_1, \mu_2, \cdots, \mu_m$, respectively, then for any Borel set B of real numbers, we have

$$\text{Ch}\{\xi \in B\} = \begin{cases} \sup\limits_{x_1, x_2 \cdots, x_m} \left(\left(\min\limits_{1 \le i \le m} \dfrac{\mu_i(x_i)}{2} \right) \wedge \sum\limits_{i=1}^{m} \{p_i \,|\, x_i \in B\} \right), \\ \qquad \text{if } \sup\limits_{x_1, x_2 \cdots, x_m} \left(\left(\min\limits_{1 \le i \le m} \dfrac{\mu_i(x_i)}{2} \right) \wedge \sum\limits_{i=1}^{m} \{p_i \,|\, x_i \in B\} \right) < 0.5 \\ 1 - \sup\limits_{x_1, x_2 \cdots, x_m} \left(\left(\min\limits_{1 \le i \le m} \dfrac{\mu_i(x_i)}{2} \right) \wedge \sum\limits_{i=1}^{m} \{p_i \,|\, x_i \in B^c\} \right), \\ \qquad \text{if } \sup\limits_{x_1, x_2 \cdots, x_m} \left(\left(\min\limits_{1 \le i \le m} \dfrac{\mu_i(x_i)}{2} \right) \wedge \sum\limits_{i=1}^{m} \{p_i \,|\, x_i \in B\} \right) \ge 0.5. \end{cases}$$

Model III

Let $\eta_1, \eta_2, \cdots, \eta_m$ be random variables, and let $u_1, u_2, \cdots, u_m$ be nonnegative numbers with $u_1 \vee u_2 \vee \cdots \vee u_m = 1$. Then

$$\xi = \begin{cases} \eta_1 \text{ with membership degree } u_1 \\ \eta_2 \text{ with membership degree } u_2 \\ \cdots \\ \eta_m \text{ with membership degree } u_m \end{cases} \tag{6.18}$$

is clearly a hybrid variable. If $\eta_1, \eta_2, \cdots, \eta_m$ have probability density functions $\phi_1, \phi_2, \cdots, \phi_m$, respectively, then for any Borel set B of real numbers, we have

$$\text{Ch}\{\xi \in B\} = \begin{cases} \max\limits_{1 \le i \le m} \left(\dfrac{u_i}{2} \wedge \displaystyle\int_B \phi_i(x) \mathrm{d}x \right), \\ \qquad \text{if } \max\limits_{1 \le i \le m} \left(\dfrac{u_i}{2} \wedge \displaystyle\int_B \phi_i(x) \mathrm{d}x \right) < 0.5 \\ 1 - \max\limits_{1 \le i \le m} \left(\dfrac{u_i}{2} \wedge \displaystyle\int_{B^c} \phi_i(x) \mathrm{d}x \right), \\ \qquad \text{if } \max\limits_{1 \le i \le m} \left(\dfrac{u_i}{2} \wedge \displaystyle\int_B \phi_i(x) \mathrm{d}x \right) \ge 0.5. \end{cases}$$

Model IV

In many statistics problems, the probability density function is completely known except for the values of one or more parameters. For example, it might be known that the lifetime ξ of a modern engine is an exponentially distributed random variable with an unknown expected value β. Usually, there is some relevant information in practice. It is thus possible to specify an interval in which the value of β is likely to lie, or to give an approximate estimate of the value of β. It is typically not possible to determine the value of β exactly. If the value of β is provided as a fuzzy variable, then ξ is a hybrid variable. More generally, suppose that ξ has a probability density function

$$\phi(x; \tilde{a}_1, \tilde{a}_2, \cdots, \tilde{a}_m), \quad x \in \Re \tag{6.19}$$

in which the parameters $\tilde{a}_1, \tilde{a}_2, \cdots, \tilde{a}_m$ are fuzzy variables rather than crisp numbers. Then ξ is a hybrid variable provided that $\phi(x; y_1, y_2, \cdots, y_m)$ is a probability density function for any $(y_1, y_2, \cdots, y_m)$ that $(\tilde{a}_1, \tilde{a}_2, \cdots, \tilde{a}_m)$ may take. If $\tilde{a}_1, \tilde{a}_2, \cdots, \tilde{a}_m$ have membership functions $\mu_1, \mu_2, \cdots, \mu_m$, respectively, then for any Borel set B of real numbers, the chance $\mathrm{Ch}\{\xi \in B\}$ is

$$\begin{cases} \sup\limits_{y_1, y_2 \cdots, y_m} \left(\left(\min\limits_{1 \le i \le m} \dfrac{\mu_i(y_i)}{2} \right) \wedge \displaystyle\int_B \phi(x; y_1, y_2, \cdots, y_m) \mathrm{d}x \right), \\ \quad \text{if } \sup\limits_{y_1, y_2, \cdots, y_m} \left(\left(\min\limits_{1 \le i \le m} \dfrac{\mu_i(y_i)}{2} \right) \wedge \displaystyle\int_B \phi(x; y_1, y_2, \cdots, y_m) \mathrm{d}x \right) < 0.5 \\ 1 - \sup\limits_{y_1, y_2 \cdots, y_m} \left(\left(\min\limits_{1 \le i \le m} \dfrac{\mu_i(y_i)}{2} \right) \wedge \displaystyle\int_{B^c} \phi(x; y_1, y_2, \cdots, y_m) \mathrm{d}x \right), \\ \quad \text{if } \sup\limits_{y_1, y_2, \cdots, y_m} \left(\left(\min\limits_{1 \le i \le m} \dfrac{\mu_i(y_i)}{2} \right) \wedge \displaystyle\int_B \phi(x; y_1, y_2, \cdots, y_m) \mathrm{d}x \right) \ge 0.5. \end{cases}$$

Hybrid Vectors

Definition 6.5. *An n-dimensional hybrid vector is a measurable function from a chance space $(\Theta, \mathcal{P}, \mathrm{Cr}) \times (\Omega, \mathcal{A}, \Pr)$ to the set of n-dimensional real vectors, i.e., for any Borel set B of $\Re^n$, the set*

$$\{\boldsymbol{\xi} \in B\} = \{(\theta, \omega) \in \Theta \times \Omega \mid \boldsymbol{\xi}(\theta, \omega) \in B\} \tag{6.20}$$

is an event.

Theorem 6.8. *The vector $(\xi_1, \xi_2, \cdots, \xi_n)$ is a hybrid vector if and only if $\xi_1, \xi_2, \cdots, \xi_n$ are hybrid variables.*

Proof: Write $\boldsymbol{\xi} = (\xi_1, \xi_2, \cdots, \xi_n)$. Suppose that $\boldsymbol{\xi}$ is a hybrid vector on the chance space $(\Theta, \mathcal{P}, \mathrm{Cr}) \times (\Omega, \mathcal{A}, \mathrm{Pr})$. For any Borel set B of $\Re$, the set $B \times \Re^{n-1}$ is a Borel set of $\Re^n$. Thus the set

$$\begin{aligned}
&\left\{(\theta, \omega) \in \Theta \times \Omega \;\middle|\; \xi_1(\theta, \omega) \in B\right\} \\
&= \left\{(\theta, \omega) \in \Theta \times \Omega \;\middle|\; \xi_1(\theta, \omega) \in B, \xi_2(\theta, \omega) \in \Re, \cdots, \xi_n(\theta, \omega) \in \Re\right\} \\
&= \left\{(\theta, \omega) \in \Theta \times \Omega \;\middle|\; \boldsymbol{\xi}(\theta, \omega) \in B \times \Re^{n-1}\right\}
\end{aligned}$$

is an event. Hence ξ_1 is a hybrid variable. A similar process may prove that $\xi_2, \xi_3, \cdots, \xi_n$ are hybrid variables.

Conversely, suppose that all $\xi_1, \xi_2, \cdots, \xi_n$ are hybrid variables on the chance space $(\Theta, \mathcal{P}, \mathrm{Cr}) \times (\Omega, \mathcal{A}, \mathrm{Pr})$. We define

$$\mathcal{B} = \left\{B \subset \Re^n \;\middle|\; \{(\theta, \omega) \in \Theta \times \Omega | \boldsymbol{\xi}(\theta, \omega) \in B\} \text{ is an event}\right\}.$$

The vector $\boldsymbol{\xi} = (\xi_1, \xi_2, \cdots, \xi_n)$ is proved to be a hybrid vector if we can prove that $\mathcal{B}$ contains all Borel sets of $\Re^n$. First, the class $\mathcal{B}$ contains all open intervals of $\Re^n$ because

$$\left\{(\theta, \omega) \;\middle|\; \boldsymbol{\xi}(\theta, \omega) \in \prod_{i=1}^{n} (a_i, b_i)\right\} = \bigcap_{i=1}^{n} \left\{(\theta, \omega) \;\middle|\; \xi_i(\theta, \omega) \in (a_i, b_i)\right\}$$

is an event. Next, the class $\mathcal{B}$ is a σ-algebra of $\Re^n$ because (i) we have $\Re^n \in \mathcal{B}$ since $\{(\theta, \omega) | \boldsymbol{\xi}(\theta, \omega) \in \Re^n\} = \Theta \times \Omega$; (ii) if $B \in \mathcal{B}$, then

$$\left\{(\theta, \omega) \in \Theta \times \Omega \;\middle|\; \boldsymbol{\xi}(\theta, \omega) \in B\right\}$$

is an event, and

$$\{(\theta, \omega) \in \Theta \times \Omega \mid \boldsymbol{\xi}(\theta, \omega) \in B^c\} = \{(\theta, \omega) \in \Theta \times \Omega \mid \boldsymbol{\xi}(\theta, \omega) \in B\}^c$$

is an event. This means that $B^c \in \mathcal{B}$; (iii) if $B_i \in \mathcal{B}$ for $i = 1, 2, \cdots$, then $\{(\theta, \omega) \in \Theta \times \Omega | \boldsymbol{\xi}(\theta, \omega) \in B_i\}$ are events and

$$\left\{(\theta, \omega) \in \Theta \times \Omega \;\middle|\; \boldsymbol{\xi}(\theta, \omega) \in \bigcup_{i=1}^{\infty} B_i\right\} = \bigcup_{i=1}^{\infty} \{(\theta, \omega) \in \Theta \times \Omega \mid \boldsymbol{\xi}(\theta, \omega) \in B_i\}$$

is an event. This means that $\cup_i B_i \in \mathcal{B}$. Since the smallest σ-algebra containing all open intervals of $\Re^n$ is just the Borel algebra of $\Re^n$, the class $\mathcal{B}$ contains all Borel sets of $\Re^n$. The theorem is proved.

Hybrid Arithmetic

Definition 6.6. *Let $f: \Re^n \to \Re$ be a measurable function, and $\xi_1, \xi_2, \cdots, \xi_n$ hybrid variables on the chance space $(\Theta, \mathcal{P}, \mathrm{Cr}) \times (\Omega, \mathcal{A}, \mathrm{Pr})$. Then $\xi = f(\xi_1, \xi_2, \cdots, \xi_n)$ is a hybrid variable defined as*

$$\xi(\theta, \omega) = f(\xi_1(\theta, \omega), \xi_2(\theta, \omega), \cdots, \xi_n(\theta, \omega)), \quad \forall(\theta, \omega) \in \Theta \times \Omega. \tag{6.21}$$

Example 6.4. Let ξ_1 and ξ_2 be two hybrid variables defined as follows,

$$\xi_1 \sim \begin{cases} \mathcal{N}(u_1, \sigma_1^2) \text{ with membership degree } 0.7 \\ \mathcal{N}(u_2, \sigma_2^2) \text{ with membership degree } 1.0, \end{cases}$$

$$\xi_2 \sim \begin{cases} \mathcal{N}(u_3, \sigma_3^2) \text{ with membership degree } 1.0 \\ \mathcal{N}(u_4, \sigma_4^2) \text{ with membership degree } 0.8. \end{cases}$$

Then the sum of ξ_1 and ξ_2 is

$$\xi \sim \begin{cases} \mathcal{N}(u_1 + u_3, \sigma_1^2 + \sigma_3^2) \text{ with membership degree } 0.7 \\ \mathcal{N}(u_1 + u_4, \sigma_1^2 + \sigma_4^2) \text{ with membership degree } 0.7 \\ \mathcal{N}(u_2 + u_3, \sigma_2^2 + \sigma_3^2) \text{ with membership degree } 1.0 \\ \mathcal{N}(u_2 + u_4, \sigma_2^2 + \sigma_4^2) \text{ with membership degree } 0.8. \end{cases}$$

Example 6.5. Let ξ_1 and ξ_2 be two hybrid variables defined as follows,

$$\xi_1 = \begin{cases} (a_1, a_2, a_3, a_4) \text{ with probability } 0.3 \\ (b_1, b_2, b_3, b_4) \text{ with probability } 0.7, \end{cases}$$

$$\xi_2 = \begin{cases} (c_1, c_2, c_3, c_4) \text{ with probability } 0.6 \\ (d_1, d_2, d_3, d_4) \text{ with probability } 0.4. \end{cases}$$

Then the sum of ξ_1 and ξ_2 is

$$\xi_1 + \xi_2 = \begin{cases} (a_1 + c_1, a_2 + c_2, a_3 + c_3, a_4 + c_4) \text{ with probability } 0.18 \\ (a_1 + d_1, a_2 + d_2, a_3 + d_3, a_4 + d_4) \text{ with probability } 0.12 \\ (b_1 + c_1, b_2 + c_2, b_3 + c_3, b_4 + c_4) \text{ with probability } 0.42 \\ (b_1 + d_1, b_2 + d_2, b_3 + d_3, b_4 + d_4) \text{ with probability } 0.28. \end{cases}$$

Expected Value

Li and Liu [161] suggested the following definition of expected value operator of hybrid variables.

Definition 6.7. *Let ξ be a hybrid variable. Then the expected value of ξ is defined by*

$$E[\xi] = \int_0^{+\infty} \text{Ch}\{\xi \geq r\}\text{d}r - \int_{-\infty}^0 \text{Ch}\{\xi \leq r\}\text{d}r \tag{6.22}$$

provided that at least one of the two integrals is finite.

Example 6.6. If a hybrid variable ξ degenerates to a random variable η, then

$$\text{Ch}\{\xi \leq x\} = \Pr\{\eta \leq x\}, \quad \text{Ch}\{\xi \geq x\} = \Pr\{\eta \geq x\}, \quad \forall x \in \Re.$$

It follows from (6.22) that $E[\xi] = E[\eta]$. In other words, the expected value operator of hybrid variable coincides with that of random variable.

Example 6.7. If a hybrid variable ξ degenerates to a fuzzy variable $\tilde{a}$, then

$$\text{Ch}\{\xi \le x\} = \text{Cr}\{\tilde{a} \le x\}, \quad \text{Ch}\{\xi \ge x\} = \text{Cr}\{\tilde{a} \ge x\}, \quad \forall x \in \Re.$$

It follows from (6.22) that $E[\xi] = E[\tilde{a}]$. In other words, the expected value operator of hybrid variable coincides with that of fuzzy variable.

Example 6.8. Let $\tilde{a}$ be a fuzzy variable and η a random variable with finite expected values. Then the hybrid variable $\xi = \tilde{a} + \eta$ has expected value $E[\xi] = E[\tilde{a}] + E[\eta]$.

Theorem 6.9. *Let ξ be a hybrid variable with finite expected values. Then for any real numbers a and b, we have*

$$E[a\xi + b] = aE[\xi] + b. \tag{6.23}$$

Proof: STEP 1: We first prove that $E[\xi + b] = E[\xi] + b$ for any real number b. If $b \ge 0$, we have

$$\begin{aligned} E[\xi + b] &= \int_0^{+\infty} \text{Ch}\{\xi + b \ge r\}\text{d}r - \int_{-\infty}^0 \text{Ch}\{\xi + b \le r\}\text{d}r \\ &= \int_0^{+\infty} \text{Ch}\{\xi \ge r - b\}\text{d}r - \int_{-\infty}^0 \text{Ch}\{\xi \le r - b\}\text{d}r \\ &= E[\xi] + \int_0^b (\text{Ch}\{\xi \ge r - b\} + \text{Ch}\{\xi < r - b\})\text{d}r \\ &= E[\xi] + b. \end{aligned}$$

If $b < 0$, then we have

$$E[a\xi + b] = E[\xi] - \int_b^0 (\text{Ch}\{\xi \ge r - b\} + \text{Ch}\{\xi < r - b\})\text{d}r = E[\xi] + b.$$

STEP 2: We prove $E[a\xi] = aE[\xi]$. If $a = 0$, then the equation $E[a\xi] = aE[\xi]$ holds trivially. If $a > 0$, we have

$$\begin{aligned} E[a\xi] &= \int_0^{+\infty} \text{Ch}\{a\xi \ge r\}\text{d}r - \int_{-\infty}^0 \text{Ch}\{a\xi \le r\}\text{d}r \\ &= \int_0^{+\infty} \text{Ch}\{\xi \ge r/a\}\text{d}r - \int_{-\infty}^0 \text{Ch}\{\xi \le r/a\}\text{d}r \\ &= a\int_0^{+\infty} \text{Ch}\{\xi \ge t\}\text{d}t - a\int_{-\infty}^0 \text{Ch}\{\xi \le t\}\text{d}t \\ &= aE[\xi]. \end{aligned}$$

If $a < 0$, we have

$$\begin{aligned} E[a\xi] &= \int_0^{+\infty} \text{Ch}\{a\xi \ge r\}\text{d}r - \int_{-\infty}^0 \text{Ch}\{a\xi \le r\}\text{d}r \\ &= \int_0^{+\infty} \text{Ch}\{\xi \le r/a\}\text{d}r - \int_{-\infty}^0 \text{Ch}\{\xi \ge r/a\}\text{d}r \\ &= a\int_0^{+\infty} \text{Ch}\{\xi \ge t\}\text{d}t - a\int_{-\infty}^0 \text{Ch}\{\xi \le t\}\text{d}t \\ &= aE[\xi]. \end{aligned}$$

STEP 3: For any real numbers a and b, it follows from Steps 1 and 2 that

$$E[a\xi + b] = E[a\xi] + b = aE[\xi] + b.$$

The theorem is proved.

Critical Values

In order to rank hybrid variables, Li and Liu [161] presented the following definition of critical values of hybrid variables.

Definition 6.8. *Let ξ be a hybrid variable, and $\alpha \in (0, 1]$. Then*

$$\xi_{\sup}(\alpha) = \sup\left\{r \mid \text{Ch}\left\{\xi \ge r\right\} \ge \alpha\right\} \tag{6.24}$$

is called the α-optimistic value to ξ, and

$$\xi_{\inf}(\alpha) = \inf\left\{r \mid \text{Ch}\left\{\xi \le r\right\} \ge \alpha\right\} \tag{6.25}$$

is called the α-pessimistic value to ξ.

The hybrid variable ξ reaches upwards of the α-optimistic value $\xi_{\sup}(\alpha)$, and is below the α-pessimistic value $\xi_{\inf}(\alpha)$ with chance α.

Example 6.9. If a hybrid variable ξ degenerates to a random variable η, then

$$\text{Ch}\{\xi \le x\} = \Pr\{\eta \le x\}, \quad \text{Ch}\{\xi \ge x\} = \Pr\{\eta \ge x\}, \quad \forall x \in \Re.$$

It follows from the definition of critical values that

$$\xi_{\sup}(\alpha) = \eta_{\sup}(\alpha), \quad \xi_{\inf}(\alpha) = \eta_{\inf}(\alpha), \quad \forall \alpha \in (0, 1].$$

In other words, the critical values of hybrid variable coincide with that of random variable.

Example 6.10. If a hybrid variable ξ degenerates to a fuzzy variable $\tilde{a}$, then

$$\text{Ch}\{\xi \le x\} = \text{Cr}\{\tilde{a} \le x\}, \quad \text{Ch}\{\xi \ge x\} = \text{Cr}\{\tilde{a} \ge x\}, \quad \forall x \in \Re.$$

It follows from the definition of critical values that

$$\xi_{\sup}(\alpha) = \tilde{a}_{\sup}(\alpha), \quad \xi_{\inf}(\alpha) = \tilde{a}_{\inf}(\alpha), \quad \forall \alpha \in (0, 1].$$

In other words, the critical values of hybrid variable coincide with that of fuzzy variable.

Theorem 6.10. *Let ξ be a hybrid variable. Then we have*
(a) $\xi_{\sup}(\alpha)$ is a decreasing and left-continuous function of α;
(b) $\xi_{\inf}(\alpha)$ is an increasing and left-continuous function of α.

Proof: (a) It is easy to prove that $\xi_{\inf}(\alpha)$ is an increasing function of α. Next, we prove the left-continuity of $\xi_{\inf}(\alpha)$ with respect to α. Let $\{\alpha_i\}$ be an arbitrary sequence of positive numbers such that $\alpha_i \uparrow \alpha$. Then $\{\xi_{\inf}(\alpha_i)\}$ is an increasing sequence. If the limitation is equal to $\xi_{\inf}(\alpha)$, then the left-continuity is proved. Otherwise, there exists a number z^* such that

$$\lim_{i \to \infty} \xi_{\inf}(\alpha_i) < z^* < \xi_{\inf}(\alpha).$$

Thus $\text{Ch}\{\xi \leq z^*\} \geq \alpha_i$ for each i. Letting $i \to \infty$, we get $\text{Ch}\{\xi \leq z^*\} \geq \alpha$. Hence $z^* \geq \xi_{\inf}(\alpha)$. A contradiction proves the left-continuity of $\xi_{\inf}(\alpha)$ with respect to α. The part (b) may be proved similarly.

Ranking Criteria

Let ξ and η be two hybrid variables. Different from the situation of real numbers, there does not exist a natural ordership in a hybrid world. Thus an important problem appearing in this area is how to rank hybrid variables. Here we give four ranking criteria.

Expected Value Criterion: We say $\xi > \eta$ if and only if $E[\xi] > E[\eta]$.

Optimistic Value Criterion: We say $\xi > \eta$ if and only if, for some predetermined confidence level $\alpha \in (0, 1]$, we have $\xi_{\sup}(\alpha) > \eta_{\sup}(\alpha)$, where $\xi_{\sup}(\alpha)$ and $\eta_{\sup}(\alpha)$ are the α-optimistic values of ξ and η, respectively.

Pessimistic Value Criterion: We say $\xi > \eta$ if and only if, for some predetermined confidence level $\alpha \in (0, 1]$, we have $\xi_{\inf}(\alpha) > \eta_{\inf}(\alpha)$, where $\xi_{\inf}(\alpha)$ and $\eta_{\inf}(\alpha)$ are the α-pessimistic values of ξ and η, respectively.

Chance Criterion: We say $\xi > \eta$ if and only if, for some predetermined levels $\overline{r}$, we have $\text{Ch}\{\xi \geq \overline{r}\} > \text{Ch}\{\eta \geq \overline{r}\}$.

6.2 Expected Value Model

In order to obtain the decision with maximum expected return subject to expected constraints, this section introduces the following hybrid EVM,

$$\begin{cases} \max E[f(\boldsymbol{x}, \boldsymbol{\xi})] \\ \text{subject to:} \\ \qquad E[g_j(\boldsymbol{x}, \boldsymbol{\xi})] \le 0, \ j = 1, 2, \cdots, p \end{cases} \tag{6.26}$$

where $\boldsymbol{x}$ is a decision vector, $\boldsymbol{\xi}$ is a hybrid vector, f is the objective function, and g_j are the constraint functions for $j = 1, 2, \cdots, p$.

In practice, a decision maker may want to optimize multiple objectives. Thus we have the following hybrid expected value multiobjective programming (EVMOP),

$$\begin{cases} \max \left[E[f_1(\boldsymbol{x}, \boldsymbol{\xi})], E[f_2(\boldsymbol{x}, \boldsymbol{\xi})], \cdots, E[f_m(\boldsymbol{x}, \boldsymbol{\xi})]\right] \\ \text{subject to:} \\ \qquad E[g_j(\boldsymbol{x}, \boldsymbol{\xi})] \le 0, \ j = 1, 2, \cdots, p \end{cases} \tag{6.27}$$

where $f_i(\boldsymbol{x}, \boldsymbol{\xi})$ are objective functions for $i = 1, 2, \cdots, m$, and $g_j(\boldsymbol{x}, \boldsymbol{\xi})$ are constraint functions for $j = 1, 2, \cdots, p$.

In order to balance the multiple conflicting objectives, a decision-maker may establish a hierarchy of importance among these incompatible goals so as to satisfy as many goals as possible in the order specified. Thus we have a hybrid expected value goal programming (EVGP),

$$\begin{cases} \min \displaystyle\sum_{j=1}^{l} P_j \sum_{i=1}^{m} (u_{ij} d_i^+ \vee 0 + v_{ij} d_i^- \vee 0) \\ \text{subject to:} \\ \qquad E[f_i(\boldsymbol{x}, \boldsymbol{\xi})] - b_i = d_i^+, \quad i = 1, 2, \cdots, m \\ \qquad b_i - E[f_i(\boldsymbol{x}, \boldsymbol{\xi})] = d_i^-, \quad i = 1, 2, \cdots, m \\ \qquad E[g_j(\boldsymbol{x}, \boldsymbol{\xi})] \le 0, \qquad\quad j = 1, 2, \cdots, p \end{cases} \tag{6.28}$$

where P_j is the preemptive priority factor which expresses the relative importance of various goals, $P_j \gg P_{j+1}$, for all j, u_{ij} is the weighting factor corresponding to positive deviation for goal i with priority j assigned, v_{ij} is the weighting factor corresponding to negative deviation for goal i with priority j assigned, $d_i^+ \vee 0$ is the positive deviation from the target of goal i, $d_i^- \vee 0$ is the negative deviation from the target of goal i, f_i is a function in goal constraints, g_j is a function in real constraints, b_i is the target value according to goal i, l is the number of priorities, m is the number of goal constraints, and p is the number of real constraints.

6.3 Chance-Constrained Programming

Assume that $\boldsymbol{x}$ is a decision vector, $\boldsymbol{\xi}$ is a hybrid vector, $f(\boldsymbol{x}, \boldsymbol{\xi})$ is a return function, and $g_j(\boldsymbol{x}, \boldsymbol{\xi})$ are constraint functions, $j = 1, 2, \cdots, p$. Since the hybrid constraints $g_j(\boldsymbol{x}, \boldsymbol{\xi}) \le 0, j = 1, 2, \cdots, p$ do not define a deterministic feasible set, it is naturally desired that the hybrid constraints hold with chance α, where α is a specified confidence level. Then we have a chance constraint as follows,

$$\text{Ch}\{g_j(\boldsymbol{x}, \boldsymbol{\xi}) \le 0, j = 1, 2, \cdots, p\} \ge \alpha. \tag{6.29}$$

Maximax Chance-Constrained Programming

If we want to maximize the optimistic value to the hybrid return subject to some chance constraints, we have the following hybrid maximax CCP,

$$\begin{cases} \max\limits_{\boldsymbol{x}} \max\limits_{\overline{f}} \overline{f} \\ \text{subject to:} \\ \qquad \text{Ch}\left\{f(\boldsymbol{x}, \boldsymbol{\xi}) \ge \overline{f}\right\} \ge \beta \\ \qquad \text{Ch}\left\{g_j(\boldsymbol{x}, \boldsymbol{\xi}) \le 0\right\} \ge \alpha_j, \quad j = 1, 2, \cdots, p \end{cases} \tag{6.30}$$

where α_j and β are specified confidence levels for $j = 1, 2, \cdots, p$, and $\max \overline{f}$ is the β-optimistic return.

In practice, we may have multiple objectives. We thus have the following hybrid maximax chance-constrained multiobjective programming (CCMOP),

$$\begin{cases} \max\limits_{\boldsymbol{x}} \left[\max\limits_{\overline{f}_1} \overline{f}_1, \max\limits_{\overline{f}_2} \overline{f}_2, \cdots, \max\limits_{\overline{f}_m} \overline{f}_m\right] \\ \text{subject to:} \\ \qquad \text{Ch}\left\{f_i(\boldsymbol{x}, \boldsymbol{\xi}) \ge \overline{f}_i\right\} \ge \beta_i, \quad i = 1, 2, \cdots, m \\ \qquad \text{Ch}\left\{g_j(\boldsymbol{x}, \boldsymbol{\xi}) \le 0\right\} \ge \alpha_j, \quad j = 1, 2, \cdots, p \end{cases} \tag{6.31}$$

where β_i are predetermined confidence levels for $i = 1, 2, \cdots, m$, and $\max \overline{f}_i$ are the β-optimistic values to the return functions $f_i(\boldsymbol{x}, \boldsymbol{\xi})$, $i = 1, 2, \cdots, m$, respectively.

If the priority structure and target levels are set by the decision-maker, then we have a minimin chance-constrained goal programming (CCGP),

$$\begin{cases} \min\limits_{\boldsymbol{x}} \sum\limits_{j=1}^{l} P_j \sum\limits_{i=1}^{m} \left(u_{ij}\left(\min\limits_{d_i^+} d_i^+ \vee 0\right) + v_{ij}\left(\min\limits_{d_i^-} d_i^- \vee 0\right)\right) \\ \text{subject to:} \\ \qquad \text{Ch}\left\{f_i(\boldsymbol{x}, \boldsymbol{\xi}) - b_i \le d_i^+\right\} \ge \beta_i^+, \quad i = 1, 2, \cdots, m \\ \qquad \text{Ch}\left\{b_i - f_i(\boldsymbol{x}, \boldsymbol{\xi}) \le d_i^-\right\} \ge \beta_i^-, \quad i = 1, 2, \cdots, m \\ \qquad \text{Ch}\left\{g_j(\boldsymbol{x}, \boldsymbol{\xi}) \le 0\right\} \ge \alpha_j, \qquad\quad j = 1, 2, \cdots, p \end{cases} \tag{6.32}$$

where P_j is the preemptive priority factor which expresses the relative importance of various goals, $P_j \gg P_{j+1}$, for all j, u_{ij} is the weighting factor corresponding to positive deviation for goal i with priority j assigned, v_{ij}

is the weighting factor corresponding to negative deviation for goal i with priority j assigned, $\min d_i^+ \vee 0$ is the β_i^+-optimistic positive deviation from the target of goal i, $\min d_i^- \vee 0$ is the β_i^--optimistic negative deviation from the target of goal i, b_i is the target value according to goal i, and l is the number of priorities.

Minimax Chance-Constrained Programming

If we want to maximize the pessimistic value subject to some chance constraints, then we have the following hybrid minimax CCP,

$$\begin{cases} \max\limits_{\boldsymbol{x}} \min\limits_{\overline{f}} \overline{f} \\ \text{subject to:} \\ \qquad \text{Ch}\left\{f(\boldsymbol{x},\boldsymbol{\xi}) \le \overline{f}\right\} \ge \beta \\ \qquad \text{Ch}\left\{g_j(\boldsymbol{x},\boldsymbol{\xi}) \le 0\right\} \ge \alpha_j, \quad j = 1,2,\cdots,p \end{cases} \tag{6.33}$$

where α_j and β are specified confidence levels for $j = 1,2,\cdots,p$, and $\min \overline{f}$ is the β-pessimistic return.

If there are multiple objectives, then we have the following hybrid minimax CCMOP,

$$\begin{cases} \max\limits_{\boldsymbol{x}} \left[\min\limits_{\overline{f}_1} \overline{f}_1,\ \min\limits_{\overline{f}_2} \overline{f}_2,\ \cdots,\ \min\limits_{\overline{f}_m} \overline{f}_m\right] \\ \text{subject to:} \\ \qquad \text{Ch}\left\{f_i(\boldsymbol{x},\boldsymbol{\xi}) \le \overline{f}_i\right\} \ge \beta_i, \quad i = 1,2,\cdots,m \\ \qquad \text{Ch}\left\{g_j(\boldsymbol{x},\boldsymbol{\xi}) \le 0\right\} \ge \alpha_j, \quad j = 1,2,\cdots,p \end{cases} \tag{6.34}$$

where $\min \overline{f}_i$ are the β_i-pessimistic values to the return functions $f_i(\boldsymbol{x},\boldsymbol{\xi})$, $i = 1,2,\cdots,m$, respectively.

We can also formulate a hybrid decision system as a hybrid minimax CCGP according to the priority structure and target levels set by the decision-maker:

$$\begin{cases} \min\limits_{\boldsymbol{x}} \sum\limits_{j=1}^{l} P_j \sum\limits_{i=1}^{m} \left[u_{ij}\left(\max\limits_{d_i^+} d_i^+ \vee 0\right) + v_{ij}\left(\max\limits_{d_i^-} d_i^- \vee 0\right)\right] \\ \text{subject to:} \\ \qquad \text{Ch}\left\{f_i(\boldsymbol{x},\boldsymbol{\xi}) - b_i \ge d_i^+\right\} \ge \beta_i^+, \quad i = 1,2,\cdots,m \\ \qquad \text{Ch}\left\{b_i - f_i(\boldsymbol{x},\boldsymbol{\xi}) \ge d_i^-\right\} \ge \beta_i^-, \quad i = 1,2,\cdots,m \\ \qquad \text{Ch}\left\{g_j(\boldsymbol{x},\boldsymbol{\xi}) \le 0\right\} \ge \alpha_j, \qquad\quad j = 1,2,\cdots,p \end{cases} \tag{6.35}$$

where P_j is the preemptive priority factor which expresses the relative importance of various goals, $P_j \gg P_{j+1}$, for all j, u_{ij} is the weighting factor corresponding to positive deviation for goal i with priority j assigned, v_{ij} is the weighting factor corresponding to negative deviation for goal i with priority j assigned, $\max d_i^+ \vee 0$ is the β_i^+-pessimistic positive deviation from the target of goal i, $\max d_i^- \vee 0$ is the β_i^--pessimistic negative deviation from the target of goal i, b_i is the target value according to goal i, and l is the number of priorities.

6.4 Dependent-Chance Programming

This section provides hybrid DCP in which the underlying philosophy is based on selecting the decision with maximum chance to meet the event. Uncertain environment and chance function are key elements in DCP. Let us redefine them in hybrid decision systems, and introduce the principle of uncertainty.

By uncertain environment (in this case the hybrid environment) we mean the hybrid constraints represented by

$$g_j(\boldsymbol{x}, \boldsymbol{\xi}) \leq 0, \quad j = 1, 2, \cdots, p \tag{6.36}$$

where $\boldsymbol{x}$ is a decision vector, and $\boldsymbol{\xi}$ is a hybrid vector. By event we mean the system of inequalities

$$h_k(\boldsymbol{x}, \boldsymbol{\xi}) \leq 0, \quad k = 1, 2, \cdots, q. \tag{6.37}$$

The chance function of an event $\mathcal{E}$ characterized by (6.37) is defined as the chance measure of the event $\mathcal{E}$, i.e.,

$$f(\boldsymbol{x}) = \text{Ch}\{h_k(\boldsymbol{x}, \boldsymbol{\xi}) \leq 0, \, k = 1, 2, \cdots, q\} \tag{6.38}$$

subject to the uncertain environment (6.36).

For each decision $\boldsymbol{x}$ and realization $\boldsymbol{\xi}$, an event $\mathcal{E}$ is said to be consistent in the uncertain environment if the following two conditions hold: (i) $h_k(\boldsymbol{x}, \boldsymbol{\xi}) \leq 0, \, k = 1, 2, \cdots, q$; and (ii) $g_j(\boldsymbol{x}, \boldsymbol{\xi}) \leq 0, \, j \in J$, where J is the index set of all dependent constraints.

Principle of Uncertainty: *The chance of a hybrid event is the chance measure value that the event is consistent in the uncertain environment.*

Assume that there are m events $\mathcal{E}_i$ characterized by $h_{ik}(\boldsymbol{x}, \boldsymbol{\xi}) \leq 0, k = 1, 2, \cdots, q_i$ for $i = 1, 2, \cdots, m$ in the uncertain environment $g_j(\boldsymbol{x}, \boldsymbol{\xi}) \leq 0, j = 1, 2, \cdots, p$. The principle of uncertainty implies that the chance function of the ith event $\mathcal{E}_i$ in the uncertain environment is

$$f_i(\boldsymbol{x}) = \text{Ch}\left\{ \begin{array}{l} h_{ik}(\boldsymbol{x}, \boldsymbol{\xi}) \leq 0, k = 1, 2, \cdots, q_i \\ g_j(\boldsymbol{x}, \boldsymbol{\xi}) \leq 0, j \in J_i \end{array} \right\} \tag{6.39}$$

where J_i are defined by

$$J_i = \left\{ j \in \{1, 2, \cdots, p\} \mid g_j(\boldsymbol{x}, \boldsymbol{\xi}) \leq 0 \text{ is a dependent constraint of } \mathcal{E}_i \right\}$$

for $i = 1, 2, \cdots, m$.

General Models

In order to maximize the chance of event in an uncertain system, we may use a hybrid DCP as follows:

$$\begin{cases} \max \text{ Ch}\{h_k(\boldsymbol{x}, \boldsymbol{\xi}) \le 0, k = 1, 2, \cdots, q\} \\ \text{subject to:} \\ \qquad g_j(\boldsymbol{x}, \boldsymbol{\xi}) \le 0, \quad j = 1, 2, \cdots, p \end{cases} \tag{6.40}$$

where $\boldsymbol{x}$ is an n-dimensional decision vector, $\boldsymbol{\xi}$ is a hybrid vector, the event $\mathcal{E}$ is characterized by $h_k(\boldsymbol{x}, \boldsymbol{\xi}) \le 0, k = 1, 2, \cdots, q$, and the uncertain environment is described by the hybrid constraints $g_j(\boldsymbol{x}, \boldsymbol{\xi}) \le 0, j = 1, 2, \cdots, p$.

If there are multiple events in the uncertain environment, then we have the following hybrid dependent-chance multiobjective programming (DCMOP),

$$\begin{cases} \max \begin{bmatrix} \text{Ch}\{h_{1k}(\boldsymbol{x}, \boldsymbol{\xi}) \le 0, k = 1, 2, \cdots, q_1\} \\ \text{Ch}\{h_{2k}(\boldsymbol{x}, \boldsymbol{\xi}) \le 0, k = 1, 2, \cdots, q_2\} \\ \cdots \\ \text{Ch}\{h_{mk}(\boldsymbol{x}, \boldsymbol{\xi}) \le 0, k = 1, 2, \cdots, q_m\} \end{bmatrix} \\ \text{subject to:} \\ \qquad g_j(\boldsymbol{x}, \boldsymbol{\xi}) \le 0, \quad j = 1, 2, \cdots, p \end{cases} \tag{6.41}$$

where the events $\mathcal{E}_i$ are characterized by $h_{ik}(\boldsymbol{x}, \boldsymbol{\xi}) \le 0, k = 1, 2, \cdots, q_i$, $i = 1, 2, \cdots, m$, respectively.

Hybrid dependent-chance goal programming (DCGP) is employed to formulate hybrid decision systems according to the priority structure and target levels set by the decision-maker,

$$\begin{cases} \min \sum\limits_{j=1}^{l} P_j \sum\limits_{i=1}^{m} (u_{ij} d_i^+ \vee 0 + v_{ij} d_i^- \vee 0) \\ \text{subject to:} \\ \quad \text{Ch}\{h_{ik}(\boldsymbol{x}, \boldsymbol{\xi}) \le 0, k = 1, 2, \cdots, q_i\} - b_i = d_i^+, \quad i = 1, 2, \cdots, m \\ \quad b_i - \text{Ch}\{h_{ik}(\boldsymbol{x}, \boldsymbol{\xi}) \le 0, k = 1, 2, \cdots, q_i\} = d_i^-, \quad i = 1, 2, \cdots, m \\ \quad g_j(\boldsymbol{x}, \boldsymbol{\xi}) \le 0, \qquad\qquad j = 1, 2, \cdots, p \end{cases}$$

where P_j is the preemptive priority factor which expresses the relative importance of various goals, $P_j \gg P_{j+1}$, for all j, u_{ij} is the weighting factor corresponding to positive deviation for goal i with priority j assigned, v_{ij} is the weighting factor corresponding to negative deviation for goal i with priority j assigned, $d_i^+ \vee 0$ is the positive deviation from the target of goal i, $d_i^- \vee 0$ is the negative deviation from the target of goal i, g_j is a function in system constraints, b_i is the target value according to goal i, l is the number of priorities, m is the number of goal constraints, and p is the number of system constraints.

6.5 Hybrid Intelligent Algorithm

Liu [189] first designed hybrid simulations to estimate the uncertain functions like

$$\begin{aligned} &U_1: \boldsymbol{x} \to E[f(\boldsymbol{x},\boldsymbol{\xi})], \\ &U_2: \boldsymbol{x} \to \mathrm{Ch}\left\{g_j(\boldsymbol{x},\boldsymbol{\xi}) \le 0, j=1,2,\cdots,p\right\}, \\ &U_3: \boldsymbol{x} \to \max\left\{\overline{f} \mid \mathrm{Ch}\left\{f(\boldsymbol{x},\boldsymbol{\xi}) \ge \overline{f}\right\} \ge \alpha\right\}. \end{aligned} \tag{6.42}$$

Hybrid Simulation for $U_1(\boldsymbol{x})$

In order to compute $U_1(\boldsymbol{x})$, we randomly generate $\theta_1,\theta_2,\cdots,\theta_N$ from the credibility space $(\Theta,\mathcal{P},\mathrm{Cr})$, and $\omega_1,\omega_2,\cdots,\omega_N$ from the probability space $(\Omega,\mathcal{A},\Pr)$. For each θ_k, we estimate the probabilities like

$$\Pr\{f(\boldsymbol{x},\boldsymbol{\xi}(\theta_k,\cdot)) \ge r\}, \quad \Pr\{f(\boldsymbol{x},\boldsymbol{\xi}(\theta_k,\cdot)) < r\}$$

by the technique of stochastic simulation via the samples $\omega_1,\omega_2,\cdots,\omega_N$. For any number $r \ge 0$, the value $\mathrm{Ch}\{f(\boldsymbol{x},\boldsymbol{\xi}) \ge r\}$ is

$$\begin{cases} \max\limits_{1\le k\le N} \mathrm{Cr}\{\theta_k\} \wedge \Pr\{f(\boldsymbol{x},\boldsymbol{\xi}(\theta_k,\cdot)) \ge r\}, \\ \qquad \text{if } \max\limits_{1\le k\le N} \mathrm{Cr}\{\theta_k\} \wedge \Pr\{f(\boldsymbol{x},\boldsymbol{\xi}(\theta_k,\cdot)) \ge r\} < 0.5 \\ 1-\max\limits_{1\le k\le N} \mathrm{Cr}\{\theta_k\} \wedge \Pr\{f(\boldsymbol{x},\boldsymbol{\xi}(\theta_k,\cdot)) < r\}, \\ \qquad \text{if } \max\limits_{1\le k\le N} \mathrm{Cr}\{\theta_k\} \wedge \Pr\{f(\boldsymbol{x},\boldsymbol{\xi}(\theta_k,\cdot)) \ge r\} \ge 0.5 \end{cases} \tag{6.43}$$

and for any number $r < 0$, the value $\mathrm{Ch}\{f(\boldsymbol{x},\boldsymbol{\xi}) \le r\}$ is

$$\begin{cases} \max\limits_{1\le k\le N} \mathrm{Cr}\{\theta_k\} \wedge \Pr\{f(\boldsymbol{x},\boldsymbol{\xi}(\theta_k,\cdot)) \le r\}, \\ \qquad \text{if } \max\limits_{1\le k\le N} \mathrm{Cr}\{\theta_k\} \wedge \Pr\{f(\boldsymbol{x},\boldsymbol{\xi}(\theta_k,\cdot)) \le r\} < 0.5 \\ 1-\max\limits_{1\le k\le N} \mathrm{Cr}\{\theta_k\} \wedge \Pr\{f(\boldsymbol{x},\boldsymbol{\xi}(\theta_k,\cdot)) > r\}, \\ \qquad \text{if } \max\limits_{1\le k\le N} \mathrm{Cr}\{\theta_k\} \wedge \Pr\{f(\boldsymbol{x},\boldsymbol{\xi}(\theta_k,\cdot)) \le r\} \ge 0.5. \end{cases} \tag{6.44}$$

The expected value is then obtained by the following process.

Algorithm 6.1 (Hybrid Simulation for $U_1(\boldsymbol{x})$)
Step 1. Set $e = 0$.
Step 2. Generate $\theta_1,\theta_2,\cdots,\theta_N$ from the credibility space $(\Theta,\mathcal{P},\mathrm{Cr})$.
Step 3. Generate $\omega_1,\omega_2,\cdots,\omega_N$ from the probability space $(\Omega,\mathcal{A},\Pr)$.
Step 4. $a = \min\limits_{1\le k\le N, 1\le j\le N} f(\boldsymbol{x},\boldsymbol{\xi}(\theta_k,\omega_j))$, $b = \max\limits_{1\le k\le N, 1\le j\le N} f(\boldsymbol{x},\boldsymbol{\xi}(\theta_k,\omega_j))$.
Step 5. Randomly generate r from $[a,b]$.
Step 6. If $r \ge 0$, then $e \leftarrow e + \mathrm{Ch}\{f(\boldsymbol{x},\boldsymbol{\xi}) \ge r\}$ by using (6.43).

Step 7. If $r < 0$, then $e \leftarrow e - \text{Ch}\{f(\boldsymbol{x}, \boldsymbol{\xi}) \le r\}$ by using (6.44).
Step 8. Repeat the fifth to seventh steps for N times.
Step 9. $E[f(\boldsymbol{\xi})] = a \vee 0 + b \wedge 0 + e \cdot (b - a)/N$.

Hybrid Simulation for $U_2(\boldsymbol{x})$

In order to compute the uncertain function $U_2(\boldsymbol{x})$, we randomly generate $\theta_1, \theta_2, \cdots, \theta_N$ from the credibility space $(\Theta, \mathcal{P}, \text{Cr})$, and $\omega_1, \omega_2, \cdots, \omega_N$ from the probability space $(\Omega, \mathcal{A}, \Pr)$. For each θ_k, we estimate

$$\Pr\{g_j(\boldsymbol{x}, \boldsymbol{\xi}(\theta_k, \cdot)) \le 0 \text{ for all } j\}, \quad \Pr\{g_j(\boldsymbol{x}, \boldsymbol{\xi}(\theta_k, \cdot)) > 0 \text{ for some } j\}$$

by stochastic simulation via the samples $\omega_1, \omega_2, \cdots, \omega_N$. Then $U_2(\boldsymbol{x})$ is

$$\begin{cases} \max\limits_{1 \le k \le N} \text{Cr}\{\theta_k\} \wedge \Pr\{g_j(\boldsymbol{x}, \boldsymbol{\xi}(\theta_k, \cdot)) \le 0 \text{ for all } j\}, \\ \qquad \text{if } \max\limits_{1 \le k \le N} \text{Cr}\{\theta_k\} \wedge \Pr\{g_j(\boldsymbol{x}, \boldsymbol{\xi}(\theta_k, \cdot)) \le 0 \text{ for all } j\} < 0.5 \\ 1 - \max\limits_{1 \le k \le N} \text{Cr}\{\theta_k\} \wedge \Pr\{g_j(\boldsymbol{x}, \boldsymbol{\xi}(\theta_k, \cdot)) > 0 \text{ for some } j\}, \\ \qquad \text{if } \max\limits_{1 \le k \le N} \text{Cr}\{\theta_k\} \wedge \Pr\{g_j(\boldsymbol{x}, \boldsymbol{\xi}(\theta_k, \cdot)) \le 0 \text{ for all } j\} \ge 0.5. \end{cases} \tag{6.45}$$

We summarize this process as follows.

Algorithm 6.2 (Hybrid Simulation for $U_2(\boldsymbol{x})$)
Step 1. Generate $\theta_1, \theta_2, \cdots, \theta_N$ from the credibility space $(\Theta, \mathcal{P}, \text{Cr})$.
Step 2. Generate $\omega_1, \omega_2, \cdots, \omega_N$ from the probability space $(\Omega, \mathcal{A}, \Pr)$.
Step 3. Employ stochastic simulation to estimate $\Pr\{g_j(\boldsymbol{x}, \boldsymbol{\xi}(\theta_k, \cdot)) \le 0 \text{ for all } j\}$ and $\Pr\{g_j(\boldsymbol{x}, \boldsymbol{\xi}(\theta_k, \cdot)) > 0 \text{ for some } j\}$ via the samples $\omega_1, \omega_2, \cdots, \omega_N$.
Step 4. Output the chance $U_2(\boldsymbol{x})$ via (6.45).

Hybrid Simulation for $U_3(\boldsymbol{x})$

In order to compute $U_3(\boldsymbol{x})$, we randomly generate $\theta_1, \theta_2, \cdots, \theta_N$ from the credibility space $(\Theta, \mathcal{P}, \text{Cr})$, and $\omega_1, \omega_2, \cdots, \omega_N$ from the probability space $(\Omega, \mathcal{A}, \Pr)$. For each θ_k, we estimate

$$\Pr\{f(\boldsymbol{x}, \boldsymbol{\xi}(\theta_k, \cdot)) \ge r\}, \quad \Pr\{f(\boldsymbol{x}, \boldsymbol{\xi}(\theta_k, \cdot)) < r\}$$

by stochastic simulation via the samples $\omega_1, \omega_2, \cdots, \omega_N$. For any number r, we set

$$L(r)=\begin{cases}\max\limits_{1\le k\le N}\mathrm{Cr}\{\theta_k\}\wedge\Pr\{f(\boldsymbol{x},\boldsymbol{\xi}(\theta_k,\cdot))\ge r\},\\ \qquad\text{if }\max\limits_{1\le k\le N}\mathrm{Cr}\{\theta_k\}\wedge\Pr\{f(\boldsymbol{x},\boldsymbol{\xi}(\theta_k,\cdot))\ge r\}<0.5\\ 1-\max\limits_{1\le k\le N}\mathrm{Cr}\{\theta_k\}\wedge\Pr\{f(\boldsymbol{x},\boldsymbol{\xi}(\theta_k,\cdot))<r\},\\ \qquad\text{if }\max\limits_{1\le k\le N}\mathrm{Cr}\{\theta_k\}\wedge\Pr\{f(\boldsymbol{x},\boldsymbol{\xi}(\theta_k,\cdot))\ge r\}\ge 0.5.\end{cases}\tag{6.46}$$

It follows from the monotonicity of $L(r)$ that we may employ bisection search to find the maximal value r such that $L(r)\ge\alpha$. This value is an estimation of $\overline{f}$. We summarize this process as follows.

Algorithm 6.3 (Hybrid Simulation for $U_3(\boldsymbol{x})$)
Step 1. Generate $\theta_1,\theta_2,\cdots,\theta_N$ from the credibility space $(\Theta,\mathcal{P},\mathrm{Cr})$.
Step 2. Generate $\omega_1,\omega_2,\cdots,\omega_N$ from the probability space $(\Omega,\mathcal{A},\Pr)$.
Step 3. $a=\min\limits_{1\le k\le N,1\le j\le N} f(\boldsymbol{x},\boldsymbol{\xi}(\theta_k,\omega_j))$, $b=\max\limits_{1\le k\le N,1\le j\le N} f(\boldsymbol{x},\boldsymbol{\xi}(\theta_k,\omega_j))$.
Step 4. Set $r=(a+b)/2$.
Step 5. Compute $L(r)$ by (6.46).
Step 6. If $L(r)>\alpha$, then set $a=r$. Otherwise, set $b=r$.
Step 7. If $|a-b|>\varepsilon$ (a predetermined precision), then go to Step 4.
Step 8. Output r as the critical value.

Hybrid Intelligent Algorithm

In order to solve hybrid programming models, we may integrate hybrid simulation, NN and GA to produce a hybrid intelligent algorithm as follows,

Algorithm 6.4 (Hybrid Intelligent Algorithm)
Step 1. Generate training input-output data for uncertain functions like

$$\begin{aligned}&U_1:\boldsymbol{x}\to E[f(\boldsymbol{x},\boldsymbol{\xi})],\\&U_2:\boldsymbol{x}\to \mathrm{Ch}\left\{g_j(\boldsymbol{x},\boldsymbol{\xi})\le 0, j=1,2,\cdots,p\right\},\\&U_3:\boldsymbol{x}\to \max\left\{\overline{f}\;\middle|\;\mathrm{Ch}\left\{f(\boldsymbol{x},\boldsymbol{\xi})\ge\overline{f}\right\}\ge\alpha\right\}\end{aligned}$$

by the hybrid simulation.
Step 2. Train a neural network to approximate the uncertain functions according to the generated training input-output data.
Step 3. Initialize *pop_size* chromosomes whose feasibility may be checked by the trained neural network.
Step 4. Update the chromosomes by crossover and mutation operations in which the feasibility of offspring may be checked by the trained neural network.

Step 5. Calculate the objective values for all chromosomes by the trained neural network.
Step 6. Compute the fitness of each chromosome according to the objective values.
Step 7. Select the chromosomes by spinning the roulette wheel.
Step 8. Repeat the fourth to seventh steps for a given number of cycles.
Step 9. Report the best chromosome as the optimal solution.

6.6 Numerical Experiments

We now provide some numerical examples to illustrate the effectiveness of hybrid intelligent algorithm.

Example 6.11. Consider the following hybrid EVM

$$\begin{cases} \min E\left[\sqrt{(x_1-\xi_1)^2+(x_2-\xi_2)^2+(x_3-\xi_3)^2}\right] \\ \text{subject to:} \\ \qquad |x_1|+|x_2|+|x_3| \le 4 \end{cases}$$

where ξ_1, ξ_2 and ξ_3 are hybrid variables defined as

$$\begin{aligned} &\xi_1 \sim \mathcal{U}(\rho-1,\rho), \text{ with } \rho=(-2,-1,0), \\ &\xi_2 \sim \mathcal{U}(\rho,\rho+1), \text{ with } \rho=(-1,0,1), \\ &\xi_3 \sim \mathcal{U}(\rho+1,\rho+2), \text{ with } \rho=(0,1,2). \end{aligned}$$

In order to solve this model, we first generate input-output data for the uncertain function

$$U: \boldsymbol{x} \to E\left[\sqrt{(x_1-\xi_1)^2+(x_2-\xi_2)^2+(x_3-\xi_3)^2}\right]$$

by hybrid simulation. Then we train an NN (3 input neurons, 5 hidden neurons, 1 output neuron) to approximate the uncertain function U. Lastly, the trained NN is embedded into a GA to produce a hybrid intelligent algorithm.

A run of the hybrid intelligent algorithm (1000 cycles in hybrid simulation, 2000 data in NN, 400 generations in GA) shows that the optimal solution is

$$\boldsymbol{x}^* = (-1.3140, 0.2198, 2.4662)$$

whose objective value is 1.5059.

Example 6.12. Let us consider the following hybrid CCP,

$$\begin{cases} \max \overline{f} \\ \text{subject to:} \\ \qquad \text{Ch}\left\{\xi_1 x_1 x_3 + \xi_2 x_2 x_4 \geq \overline{f}\right\} \geq 0.90 \\ \qquad \text{Ch}\left\{(\xi_3 + x_1 + x_2)(\xi_4 + x_3 + x_4) \leq 30\right\} \geq 0.85 \\ \qquad x_1, x_2, x_3, x_4 \geq 0 \end{cases}$$

where $\xi_1, \xi_2, \xi_3, \xi_4$ are hybrid variables defined as

$$\begin{aligned} \xi_1 &\sim \mathcal{N}(\rho, 1), \text{ with } \mu_\rho(x) = [1 - (x-1)^2] \vee 0, \\ \xi_2 &\sim \mathcal{N}(\rho, 1), \text{ with } \mu_\rho(x) = [1 - (x-2)^2] \vee 0, \\ \xi_3 &\sim \mathcal{N}(\rho, 1), \text{ with } \mu_\rho(x) = [1 - (x-3)^2] \vee 0, \\ \xi_4 &\sim \mathcal{N}(\rho, 1), \text{ with } \mu_\rho(x) = [1 - (x-4)^2] \vee 0. \end{aligned}$$

In order to solve this model, we produce input-output data for the uncertain function $U : \boldsymbol{x} \to (U_1(\boldsymbol{x}), U_2(\boldsymbol{x}))$, where

$$\begin{aligned} U_1(\boldsymbol{x}) &= \max\left\{\overline{f} \mid \text{Ch}\left\{\xi_1 x_1 x_2 + \xi_2 x_3 x_4 \geq \overline{f}\right\} \geq 0.90\right\}, \\ U_2(\boldsymbol{x}) &= \text{Ch}\left\{(\xi_3 + x_1 + x_2)(\xi_4 + x_3 + x_4) \leq 30\right\}, \end{aligned}$$

by the hybrid simulation. Based on the input-output data, we train an NN (4 input neurons, 8 hidden neurons, 2 output neurons) to approximate the uncertain function U. After that, the trained NN is embedded into a GA to produce a hybrid intelligent algorithm.

A run of the hybrid intelligent algorithm (5000 cycles in simulation, 3000 training data in NN, 600 generations in GA) shows that the optimal solution is

$$(x_1^*, x_2^*, x_3^*, x_4^*) = (0.000, 1.303, 0.000, 1.978)$$

whose objective value is 2.85. Moreover, we have

$$\begin{aligned} &\text{Ch}\left\{\xi_1 x_1^* x_3^* + \xi_2 x_2^* x_4^* \geq 2.85\right\} \approx 0.90, \\ &\text{Ch}\left\{(\xi_3 + x_1^* + x_2^*)(\xi_4 + x_3^* + x_4^*) \leq 30\right\} \approx 0.85. \end{aligned}$$

Example 6.13. This is a hybrid DCP which maximizes the chance of an uncertain event subject to a deterministic constraint,

$$\begin{cases} \max \text{Ch}\left\{\xi_1 x_1 + \xi_2 x_2 + \xi_3 x_3 \geq 5\right\} \\ \text{subject to:} \\ \qquad x_1^2 + x_2^2 + x_3^2 \leq 4 \end{cases}$$

where ξ_1, ξ_2, ξ_3 are hybrid variables defined as

$$\begin{aligned}
\xi_1 &\sim \mathcal{N}(\rho, 1), \text{ with } \mu_\rho(x) = 1/[1 + (x-1)^2], \\
\xi_2 &\sim \mathcal{N}(\rho, 1), \text{ with } \mu_\rho(x) = 1/[1 + (x-2)^2], \\
\xi_3 &\sim \mathcal{N}(\rho, 1), \text{ with } \mu_\rho(x) = 1/[1 + (x-3)^2].
\end{aligned}$$

We produce a set of input-output data for the uncertain function

$$U : (x_1, x_2, x_3) \to \text{Ch}\left\{\xi_1 x_1 + \xi_2 x_2 + \xi_3 x_3 \geq 5\right\}$$

by the hybrid simulation. According to the generated data, we train a feed-forward NN (3 input neurons, 5 hidden neurons, 1 output neuron) to approximate the uncertain function U. Then we integrate the trained NN and GA to produce a hybrid intelligent algorithm.

A run of the hybrid intelligent algorithm (6000 cycles in simulation, 2000 data in NN, 500 generations in GA) shows that the optimal solution is

$$(x_1^*, x_2^*, x_3^*) = (0.6847, 1.2624, 1.3919)$$

whose chance is 0.9.

Example 6.14. We consider the following hybrid DCGP,

$$\begin{cases}
\text{lexmin}\left\{d_1^- \vee 0, d_2^- \vee 0, d_3^- \vee 0\right\} \\
\text{subject to:} \\
\qquad 0.88 - \text{Ch}\left\{x_1 + x_5 = 1\right\} = d_1^- \\
\qquad 0.85 - \text{Ch}\left\{x_2 + x_3 = 2\right\} = d_2^- \\
\qquad 0.82 - \text{Ch}\left\{x_4 + x_6 = 3\right\} = d_3^- \\
\qquad x_1^2 \leq \xi_1 \\
\qquad x_2^2 + x_3^2 + x_4^2 \leq \xi_2 \\
\qquad x_5^2 + x_6^2 \leq \xi_3
\end{cases}$$

where $\xi_1, \xi_2, \xi_3, \xi_4$ are hybrid variables defined as follows,

$$\begin{aligned}
\xi_1 &\sim \mathcal{EXP}(\rho), \text{ with } \mu_\rho(x) = [1 - (x-6)^2] \vee 0, \\
\xi_2 &\sim \mathcal{EXP}(\rho), \text{ with } \mu_\rho(x) = [1 - (x-30)^2] \vee 0, \\
\xi_3 &\sim \mathcal{EXP}(\rho), \text{ with } \mu_\rho(x) = [1 - (x-18)^2] \vee 0.
\end{aligned}$$

At the first priority level, there is one event denoted by $\mathcal{E}_1$, which will be fulfilled by $x_1 + x_5 = 1$. It is clear that the support $\mathcal{E}_1^* = \{x_1, x_5\}$ and the dependent support $\mathcal{E}_1^{**} = \{x_1, x_5, x_6\}$. It follows from the principle of uncertainty that the chance function of the event $\mathcal{E}_1$ is

$$f_1(\boldsymbol{x}) = \text{Ch}\left\{\begin{array}{l} x_1 + x_5 = 1 \\ x_1^2 \leq \xi_1 \\ x_5^2 + x_6^2 \leq \xi_3 \end{array}\right\}.$$

At the second priority level, there is an event $\mathcal{E}_2$ which will be fulfilled by $x_2 + x_3 = 2$. The support $\mathcal{E}_2^* = \{x_2, x_3\}$ and the dependent support $\mathcal{E}_2^{**} = \{x_2, x_3, x_4\}$. It follows from the principle of uncertainty that the chance function of the event $\mathcal{E}_2$ is

$$f_2(\boldsymbol{x}) = \text{Ch}\left\{\begin{array}{l} x_2 + x_3 = 2 \\ x_2^2 + x_3^2 + x_4^2 \le \xi_2 \end{array}\right\}.$$

At the third priority level, there is an event $\mathcal{E}_3$ which will be fulfilled by $x_4 + x_6 = 3$. The support $\mathcal{E}_3^* = \{x_4, x_6\}$ and the dependent support $\mathcal{E}_3^{**} = \{x_2, x_3, x_4, x_5, x_6\}$. It follows from the principle of uncertainty that the chance function of the event $\mathcal{E}_3$ is

$$f_3(\boldsymbol{x}) = \text{Ch}\left\{\begin{array}{l} x_4 + x_6 = 3 \\ x_2^2 + x_3^2 + x_4^2 \le \xi_2 \\ x_5^2 + x_6^2 \le \xi_3 \end{array}\right\}.$$

In order to solve the hybrid DCGP model, we encode a solution by a chromosome $V = (v_1, v_2, v_3)$. Thus a chromosome can be converted into a solution by

$$x_1 = v_1, \quad x_2 = v_2, \quad x_3 = 2 - v_2, \quad x_4 = v_3, \quad x_5 = 1 - v_1, \quad x_6 = 3 - v_3.$$

We generate a set of input-output data for the uncertain function $U : (v_1, v_2, v_3) \to (f_1(\boldsymbol{x}), f_2(\boldsymbol{x}), f_3(\boldsymbol{x}))$ by the hybrid simulation. Then we train a feedforward NN to approximate the uncertain function U. After that, the trained NN is embedded into a GA to produce a hybrid intelligent algorithm. A run of the hybrid intelligent algorithm (6000 cycles in simulation, 3000 data in NN, 1000 generations in GA) shows that the optimal solution is

$$\boldsymbol{x}^* = (0.4005, 1.0495, 0.9505, 1.7574, 0.5995, 1.2427)$$

which can satisfy the first two goals, but the third objective is 0.05. In fact, we also have

$$f_1(\boldsymbol{x}^*) \approx 0.88, \quad f_2(\boldsymbol{x}^*) \approx 0.85, \quad f_3(\boldsymbol{x}^*) \approx 0.77.$$

Chapter 7
Uncertain Programming

Uncertainty theory, founded by Liu [189] in 2007, is a branch of mathematics based on normality, monotonicity, self-duality, and countable subadditivity axioms. By uncertain programming we mean the optimization theory in generally uncertain environments. This chapter introduces the concept of uncertain variable and provides a general framework of uncertain programming.

7.1 Uncertain Variables

Let Γ be a nonempty set, and $\mathcal{L}$ a σ-algebra over Γ. Each element $\Lambda \in \mathcal{L}$ is called an *event*. In order to present an axiomatic definition of uncertain measure, it is necessary to assign to each event Λ a number $\mathcal{M}\{\Lambda\}$ which indicates the level that Λ will occur. In order to ensure that the number $\mathcal{M}\{\Lambda\}$ has certain mathematical properties, Liu [189] presented the following four axioms:

Axiom 1. *(Normality)* $\mathcal{M}\{\Gamma\} = 1$.

Axiom 2. *(Monotonicity)* $\mathcal{M}\{\Lambda_1\} \le \mathcal{M}\{\Lambda_2\}$ *whenever* $\Lambda_1 \subset \Lambda_2$.

Axiom 3. *(Self-Duality)* $\mathcal{M}\{\Lambda\} + \mathcal{M}\{\Lambda^c\} = 1$ *for any event* Λ.

Axiom 4. *(Countable Subadditivity) For every countable sequence of events* $\{\Lambda_i\}$, *we have*

$$\mathcal{M}\left\{\bigcup_{i=1}^{\infty} \Lambda_i\right\} \le \sum_{i=1}^{\infty} \mathcal{M}\{\Lambda_i\}. \tag{7.1}$$

Definition 7.1. *(Liu [189]) The set function* $\mathcal{M}$ *is called an uncertain measure if it satisfies the normality, monotonicity, self-duality, and countable subadditivity axioms.*

Example 7.1. Probability measure, credibility measure and chance measure are instances of uncertain measure.

B. Liu: Theory and Practice of Uncertain Programming, STUDFUZZ 239, pp. 111–128.
springerlink.com

Example 7.2. Let $\Gamma = \{\gamma_1, \gamma_2, \gamma_3\}$. For this case, there are only 8 events. Define

$$\mathcal{M}\{\gamma_1\} = 0.6, \quad \mathcal{M}\{\gamma_2\} = 0.3, \quad \mathcal{M}\{\gamma_3\} = 0.2,$$
$$\mathcal{M}\{\gamma_1, \gamma_2\} = 0.8, \quad \mathcal{M}\{\gamma_1, \gamma_3\} = 0.7, \quad \mathcal{M}\{\gamma_2, \gamma_3\} = 0.4,$$
$$\mathcal{M}\{\emptyset\} = 0, \quad \mathcal{M}\{\Gamma\} = 1.$$

It is clear that the set function $\mathcal{M}$ is neither probability measure nor credibility measure. However, $\mathcal{M}$ is an uncertain measure because it satisfies the four axioms.

Example 7.3. (Liu [195]) Suppose $g(x)$ is a nonnegative and integrable function on $\Re$ such that $\int_{\Re} g(x)\mathrm{d}x \geq 1$. Then for any Borel set Λ, the set function

$$\mathcal{M}\{\Lambda\} = \begin{cases} \displaystyle\int_{\Lambda} g(x)\mathrm{d}x, & \text{if } \displaystyle\int_{\Lambda} g(x)\mathrm{d}x < 0.5 \\ 1 - \displaystyle\int_{\Lambda^c} g(x)\mathrm{d}x, & \text{if } \displaystyle\int_{\Lambda^c} g(x)\mathrm{d}x < 0.5 \\ 0.5, & \text{otherwise} \end{cases} \tag{7.2}$$

is an uncertain measure on $\Re$. If $\int_{\Re} g(x)\mathrm{d}x = 1$, then $\mathcal{M}$ is just a probability measure on $\Re$.

Example 7.4. (Liu [191]) Suppose $f(x)$ is a nonnegative function and $g(x)$ is a nonnegative and integrable function satisfying

$$\sup_{x\in\Re} f(x) + \int_{\Re} g(x)\mathrm{d}x = 1. \tag{7.3}$$

Then for any Borel set Λ of real numbers, the set function

$$\mathcal{M}\{\Lambda\} = \frac{1}{2}\left(\sup_{x\in\Lambda} f(x) + \sup_{x\in\Re} f(x) - \sup_{x\in\Lambda^c} f(x)\right) + \int_{\Lambda} g(x)\mathrm{d}x \tag{7.4}$$

is an uncertain measure on $\Re$.

Theorem 7.1. *Suppose that $\mathcal{M}$ is an uncertain measure. Then $\mathcal{M}\{\emptyset\} = 0$ and $0 \leq \mathcal{M}\{\Lambda\} \leq 1$ for any event Λ.*

Proof: It follows from Axioms 1 and 3 that $\mathcal{M}\{\emptyset\} = 1 - \mathcal{M}\{\Gamma\} = 1 - 1 = 0$. It follows from Axiom 2 that $0 \leq \mathcal{M}\{\Lambda\} \leq 1$ because of $\emptyset \subset \Lambda \subset \Gamma$.

Theorem 7.2. *Suppose that $\mathcal{M}$ is an uncertain measure. Then for any events Λ_1 and Λ_2, we have*

$$\mathcal{M}\{\Lambda_1\} \vee \mathcal{M}\{\Lambda_2\} \leq \mathcal{M}\{\Lambda_1 \cup \Lambda_2\} \leq \mathcal{M}\{\Lambda_1\} + \mathcal{M}\{\Lambda_2\}. \tag{7.5}$$

Proof: The left-hand inequality follows from the monotonicity axiom and the right-hand inequality follows from the countable subadditivity axiom immediately.

Theorem 7.3. *Suppose that* $\mathcal{M}$ *is an uncertain measure. Then for any events* Λ_1 *and* Λ_2*, we have*

$$\mathcal{M}\{\Lambda_1\} + \mathcal{M}\{\Lambda_2\} - 1 \le \mathcal{M}\{\Lambda_1 \cap \Lambda_2\} \le \mathcal{M}\{\Lambda_1\} \wedge \mathcal{M}\{\Lambda_2\}. \qquad (7.6)$$

Proof: The right-hand inequality follows from the monotonicity axiom and the left-hand inequality follows from the self-duality and countable subadditivity axioms, i.e.,

$$\begin{aligned}\mathcal{M}\{\lambda_1 \cap \Lambda_2\} &= 1 - \mathcal{M}\{(\Lambda_1 \cap \Lambda_2)^c\} = 1 - \mathcal{M}\{\Lambda_1^c \cup \Lambda_2^c\} \\ &\ge 1 - (\mathcal{M}\{\Lambda_1^c\} + \mathcal{M}\{\Lambda_2^c\}) \\ &= 1 - (1 - \mathcal{M}\{\Lambda_1\}) - (1 - \mathcal{M}\{\Lambda_2\}) \\ &= \mathcal{M}\{\Lambda_1\} + \mathcal{M}\{\Lambda_2\} - 1.\end{aligned}$$

The inequalities are verified.

Definition 7.2. *(Liu [189]) Let* Γ *be a nonempty set,* $\mathcal{L}$ *a* σ*-algebra over* Γ*, and* $\mathcal{M}$ *an uncertain measure. Then the triplet* $(\Gamma, \mathcal{L}, \mathcal{M})$ *is called an uncertainty space.*

Definition 7.3. *(Liu [189]) An uncertain variable is a measurable function from an uncertainty space* $(\Gamma, \mathcal{L}, \mathcal{M})$ *to the set of real numbers, i.e., for any Borel set* B *of real numbers, we have*

$$\{\xi \in B\} = \{\gamma \in \Gamma \mid \xi(\gamma) \in B\} \in \mathcal{L}. \qquad (7.7)$$

Example 7.5. Random variable, fuzzy variable and hybrid variable are instances of uncertain variable.

Definition 7.4. *An* n*-dimensional uncertain vector is a measurable function from an uncertainty space* $(\Gamma, \mathcal{L}, \mathcal{M})$ *to the set of* n*-dimensional real vectors, i.e., for any Borel set* B *of* $\Re^n$*, the set*

$$\{\boldsymbol{\xi} \in B\} = \{\gamma \in \Gamma \mid \boldsymbol{\xi}(\gamma) \in B\} \qquad (7.8)$$

is an event.

Theorem 7.4. *The vector* $(\xi_1, \xi_2, \cdots, \xi_n)$ *is an uncertain vector if and only if* $\xi_1, \xi_2, \cdots, \xi_n$ *are uncertain variables.*

Proof: Write $\boldsymbol{\xi} = (\xi_1, \xi_2, \cdots, \xi_n)$. Suppose that $\boldsymbol{\xi}$ is an uncertain vector on the uncertainty space $(\Gamma, \mathcal{L}, \mathcal{M})$. For any Borel set B of $\Re$, the set $B \times \Re^{n-1}$ is a Borel set of $\Re^n$. Thus the set

$$\begin{aligned}&\{\gamma \in \Gamma \mid \xi_1(\gamma) \in B\}\\&= \{\gamma \in \Gamma \mid \xi_1(\gamma) \in B, \xi_2(\gamma) \in \Re, \cdots, \xi_n(\gamma) \in \Re\}\\&= \left\{\gamma \in \Gamma \mid \boldsymbol{\xi}(\gamma) \in B \times \Re^{n-1}\right\}\end{aligned}$$

is an event. Hence ξ_1 is an uncertain variable. A similar process may prove that $\xi_2, \xi_3, \cdots, \xi_n$ are uncertain variables. Conversely, suppose that all $\xi_1, \xi_2, \cdots, \xi_n$ are uncertain variables on the uncertainty space $(\Gamma, \mathcal{L}, \mathcal{M})$. We define

$$\mathcal{B} = \left\{B \subset \Re^n \mid \{\gamma \in \Gamma | \boldsymbol{\xi}(\gamma) \in B\} \text{ is an event}\right\}.$$

The vector $\boldsymbol{\xi} = (\xi_1, \xi_2, \cdots, \xi_n)$ is proved to be an uncertain vector if we can prove that $\mathcal{B}$ contains all Borel sets of $\Re^n$. First, the class $\mathcal{B}$ contains all open intervals of $\Re^n$ because

$$\left\{\gamma \mid \boldsymbol{\xi}(\gamma) \in \prod_{i=1}^{n}(a_i, b_i)\right\} = \bigcap_{i=1}^{n}\{\gamma \mid \xi_i(\gamma) \in (a_i, b_i)\}$$

is an event. Next, the class $\mathcal{B}$ is a σ-algebra of $\Re^n$ because (i) we have $\Re^n \in \mathcal{B}$ since $\{\gamma|\boldsymbol{\xi}(\gamma) \in \Re^n\} = \Gamma$; (ii) if $B \in \mathcal{B}$, then

$$\{\gamma \in \Gamma \mid \boldsymbol{\xi}(\gamma) \in B\}$$

is an event, and

$$\{\gamma \in \Gamma \mid \boldsymbol{\xi}(\gamma) \in B^c\} = \{\gamma \in \Gamma \mid \boldsymbol{\xi}(\gamma) \in B\}^c$$

is an event. This means that $B^c \in \mathcal{B}$; (iii) if $B_i \in \mathcal{B}$ for $i = 1, 2, \cdots$, then $\{\gamma \in \Gamma|\boldsymbol{\xi}(\gamma) \in B_i\}$ are events and

$$\left\{\gamma \in \Gamma \mid \boldsymbol{\xi}(\gamma) \in \bigcup_{i=1}^{\infty} B_i\right\} = \bigcup_{i=1}^{\infty}\{\gamma \in \Gamma \mid \boldsymbol{\xi}(\gamma) \in B_i\}$$

is an event. This means that $\cup_i B_i \in \mathcal{B}$. Since the smallest σ-algebra containing all open intervals of $\Re^n$ is just the Borel algebra of $\Re^n$, the class $\mathcal{B}$ contains all Borel sets of $\Re^n$. The theorem is proved.

Identification Function

A random variable may be characterized by a probability density function, and a fuzzy variable may be described by a membership function. This section will introduce an *identification function* to characterize an uncertain variable.

Definition 7.5. *(Liu [191]) An uncertain variable ξ is said to have an identification function (λ, ρ) if*

(i) $\lambda(x)$ is a nonnegative function and $\rho(x)$ is a nonnegative and integrable function such that

$$\sup_{x\in\Re}\lambda(x)+\int_{\Re}\rho(x)\mathrm{d}x=1; \tag{7.9}$$

(ii) for any Borel set B of real numbers, we have

$$\mathcal{M}\{\xi\in B\}=\frac{1}{2}\left(\sup_{x\in B}\lambda(x)+\sup_{x\in\Re}\lambda(x)-\sup_{x\in B^c}\lambda(x)\right)+\int_B\rho(x)\mathrm{d}x. \tag{7.10}$$

Some uncertain variables do not have their own identification functions. In other words, it is not true that every uncertain variable may be represented by an appropriate identification function.

Remark 7.1. The uncertain variable with identification function (λ,ρ) is essentially a fuzzy variable if

$$\sup_{x\in\Re}\lambda(x)=1.$$

For this case, λ is a membership function and $\rho\equiv 0$, a.e.

Remark 7.2. The uncertain variable with identification function (λ,ρ) is essentially a random variable if

$$\int_{\Re}\rho(x)\mathrm{d}x=1.$$

For this case, ρ is a probability density function and $\lambda\equiv 0$.

Remark 7.3. Let ξ be an uncertain variable with identification function (λ,ρ). If $\lambda(x)$ is a continuous function, then we have

$$\mathcal{M}\{\xi=x\}=\frac{\lambda(x)}{2},\quad \forall x\in\Re. \tag{7.11}$$

Remark 7.4. Let μ be a membership function, and ϕ a probability density function. Then for each $\alpha\in[0,1]$, the function (λ,ρ) with

$$\lambda=(1-\alpha)\mu,\quad \rho=\alpha\phi \tag{7.12}$$

is an identification function of some uncertain variable.

Uncertain Arithmetic

Definition 7.6. *Suppose that $f:\Re^n\to\Re$ is a measurable function, and $\xi_1,\xi_2,\cdots,\xi_n$ uncertain variables on the uncertainty space $(\Gamma,\mathcal{L},\mathcal{M})$. Then $\xi=f(\xi_1,\xi_2,\cdots,\xi_n)$ is an uncertain variable defined as*

$$\xi(\gamma)=f(\xi_1(\gamma),\xi_2(\gamma),\cdots,\xi_n(\gamma)),\quad \forall\gamma\in\Gamma. \tag{7.13}$$

The reader may wonder whether $\xi(\gamma_1, \gamma_2, \cdots, \gamma_n)$ defined by (7.13) is an uncertain variable. The following theorem answers this question.

Theorem 7.5. *Let $\boldsymbol{\xi}$ be an n-dimensional uncertain vector, and $f : \Re^n \to \Re$ a measurable function. Then $f(\boldsymbol{\xi})$ is an uncertain variable.*

Proof: Assume that $\boldsymbol{\xi}$ is an uncertain vector on the uncertainty space $(\Gamma, \mathcal{L}, \mathcal{M})$. For any Borel set B of $\Re$, since f is a measurable function, the $f^{-1}(B)$ is a Borel set of $\Re^n$. Thus the set

$$\{\gamma \in \Gamma \mid f(\boldsymbol{\xi}(\gamma)) \in B\} = \{\gamma \in \Gamma \mid \boldsymbol{\xi}(\gamma) \in f^{-1}(B)\}$$

is an event for any Borel set B. Hence $f(\boldsymbol{\xi})$ is an uncertain variable.

Expected Value

Expected value is the average value of uncertain variable in the sense of uncertain measure, and represents the size of uncertain variable.

Definition 7.7. *(Liu [189]) Let ξ be an uncertain variable. Then the expected value of ξ is defined by*

$$E[\xi] = \int_0^{+\infty} \mathcal{M}\{\xi \geq r\}\mathrm{d}r - \int_{-\infty}^0 \mathcal{M}\{\xi \leq r\}\mathrm{d}r \tag{7.14}$$

provided that at least one of the two integrals is finite.

Theorem 7.6. *Let ξ be an uncertain variable with finite expected value. Then for any real numbers a and b, we have*

$$E[a\xi + b] = aE[\xi] + b. \tag{7.15}$$

Proof: STEP 1: We first prove that $E[\xi + b] = E[\xi] + b$ for any real number b. If $b \geq 0$, we have

$$\begin{aligned}
E[\xi + b] &= \int_0^{+\infty} \mathcal{M}\{\xi + b \geq r\}\mathrm{d}r - \int_{-\infty}^0 \mathcal{M}\{\xi + b \leq r\}\mathrm{d}r \\
&= \int_0^{+\infty} \mathcal{M}\{\xi \geq r - b\}\mathrm{d}r - \int_{-\infty}^0 \mathcal{M}\{\xi \leq r - b\}\mathrm{d}r \\
&= E[\xi] + \int_0^b (\mathcal{M}\{\xi \geq r - b\} + \mathcal{M}\{\xi < r - b\})\mathrm{d}r \\
&= E[\xi] + b.
\end{aligned}$$

If $b < 0$, then we have

$$E[a\xi + b] = E[\xi] - \int_b^0 (\mathcal{M}\{\xi \geq r - b\} + \mathcal{M}\{\xi < r - b\})\mathrm{d}r = E[\xi] + b.$$

Step 2: We prove $E[a\xi] = aE[\xi]$. If $a = 0$, then the equation $E[a\xi] = aE[\xi]$ holds trivially. If $a > 0$, we have

$$\begin{aligned}
E[a\xi] &= \int_0^{+\infty} \mathcal{M}\{a\xi \ge r\}\mathrm{d}r - \int_{-\infty}^0 \mathcal{M}\{a\xi \le r\}\mathrm{d}r \\
&= \int_0^{+\infty} \mathcal{M}\{\xi \ge r/a\}\mathrm{d}r - \int_{-\infty}^0 \mathcal{M}\{\xi \le r/a\}\mathrm{d}r \\
&= a\int_0^{+\infty} \mathcal{M}\{\xi \ge t\}\mathrm{d}t - a\int_{-\infty}^0 \mathcal{M}\{\xi \le t\}\mathrm{d}t \\
&= aE[\xi].
\end{aligned}$$

If $a < 0$, we have

$$\begin{aligned}
E[a\xi] &= \int_0^{+\infty} \mathcal{M}\{a\xi \ge r\}\mathrm{d}r - \int_{-\infty}^0 \mathcal{M}\{a\xi \le r\}\mathrm{d}r \\
&= \int_0^{+\infty} \mathcal{M}\{\xi \le r/a\}\mathrm{d}r - \int_{-\infty}^0 \mathcal{M}\{\xi \ge r/a\}\mathrm{d}r \\
&= a\int_0^{+\infty} \mathcal{M}\{\xi \ge t\}\mathrm{d}t - a\int_{-\infty}^0 \mathcal{M}\{\xi \le t\}\mathrm{d}t \\
&= aE[\xi].
\end{aligned}$$

Step 3: For any real numbers a and b, it follows from Steps 1 and 2 that

$$E[a\xi + b] = E[a\xi] + b = aE[\xi] + b.$$

The theorem is proved.

Critical Values

Definition 7.8. *(Liu [189]) Let ξ be an uncertain variable, and $\alpha \in (0, 1]$. Then*

$$\xi_{\sup}(\alpha) = \sup\left\{r \;\middle|\; \mathcal{M}\{\xi \ge r\} \ge \alpha\right\} \tag{7.16}$$

is called the α-optimistic value to ξ, and

$$\xi_{\inf}(\alpha) = \inf\left\{r \;\middle|\; \mathcal{M}\{\xi \le r\} \ge \alpha\right\} \tag{7.17}$$

is called the α-pessimistic value to ξ.

Theorem 7.7. *Let ξ be an uncertain variable. Then $\xi_{\sup}(\alpha)$ is a decreasing function of α, and $\xi_{\inf}(\alpha)$ is an increasing function of α.*

Proof: It follows from the definition immediately.

Ranking Criteria

Let ξ and η be two uncertain variables. Different from the situation of real numbers, there does not exist a natural ordership in an uncertain world. Thus an important problem appearing in this area is how to rank uncertain variables. Here we give four ranking criteria.

Expected Value Criterion: We say $\xi > \eta$ if and only if $E[\xi] > E[\eta]$.

Optimistic Value Criterion: We say $\xi > \eta$ if and only if, for some predetermined confidence level $\alpha \in (0,1]$, we have $\xi_{\sup}(\alpha) > \eta_{\sup}(\alpha)$, where $\xi_{\sup}(\alpha)$ and $\eta_{\sup}(\alpha)$ are the α-optimistic values of ξ and η, respectively.

Pessimistic Value Criterion: We say $\xi > \eta$ if and only if, for some predetermined confidence level $\alpha \in (0,1]$, we have $\xi_{\inf}(\alpha) > \eta_{\inf}(\alpha)$, where $\xi_{\inf}(\alpha)$ and $\eta_{\inf}(\alpha)$ are the α-pessimistic values of ξ and η, respectively.

Chance Criterion: We say $\xi > \eta$ if and only if, for some predetermined levels $\bar{r}$, we have $\mathcal{M}\{\xi \geq \bar{r}\} > \mathcal{M}\{\eta \geq \bar{r}\}$.

7.2 Expected Value Model

In order to obtain the decision with maximum expected return subject to expected constraints, we have the following uncertain EVM,

$$\begin{cases} \max E[f(\boldsymbol{x}, \boldsymbol{\xi})] \\ \text{subject to:} \\ \qquad E[g_j(\boldsymbol{x}, \boldsymbol{\xi})] \leq 0, \; j = 1, 2, \cdots, p \end{cases} \tag{7.18}$$

where $\boldsymbol{x}$ is a decision vector, $\boldsymbol{\xi}$ is a uncertain vector, f is the objective function, and g_j are the constraint functions for $j = 1, 2, \cdots, p$.

In practice, a decision maker may want to optimize multiple objectives. Thus we have the following uncertain expected value multiobjective programming (EVMOP),

$$\begin{cases} \max\, [E[f_1(\boldsymbol{x}, \boldsymbol{\xi})], E[f_2(\boldsymbol{x}, \boldsymbol{\xi})], \cdots, E[f_m(\boldsymbol{x}, \boldsymbol{\xi})]] \\ \text{subject to:} \\ \qquad E[g_j(\boldsymbol{x}, \boldsymbol{\xi})] \leq 0, \; j = 1, 2, \cdots, p \end{cases} \tag{7.19}$$

where $f_i(\boldsymbol{x}, \boldsymbol{\xi})$ are objective functions for $i = 1, 2, \cdots, m$, and $g_j(\boldsymbol{x}, \boldsymbol{\xi})$ are constraint functions for $j = 1, 2, \cdots, p$.

In order to balance the multiple conflicting objectives, a decision-maker may establish a hierarchy of importance among these incompatible goals so as to satisfy as many goals as possible in the order specified. Thus we have an uncertain expected value goal programming (EVGP),

$$
\begin{cases}
\min \sum_{j=1}^{l} P_j \sum_{i=1}^{m} (u_{ij} d_i^+ \vee 0 + v_{ij} d_i^- \vee 0) \\
\text{subject to:} \\
\qquad E[f_i(\boldsymbol{x}, \boldsymbol{\xi})] - b_i = d_i^+, \quad i = 1, 2, \cdots, m \\
\qquad b_i - E[f_i(\boldsymbol{x}, \boldsymbol{\xi})] = d_i^-, \quad i = 1, 2, \cdots, m \\
\qquad E[g_j(\boldsymbol{x}, \boldsymbol{\xi})] \le 0, \qquad\quad j = 1, 2, \cdots, p
\end{cases}
\tag{7.20}
$$

where P_j is the preemptive priority factor which expresses the relative importance of various goals, $P_j \gg P_{j+1}$, for all j, u_{ij} is the weighting factor corresponding to positive deviation for goal i with priority j assigned, v_{ij} is the weighting factor corresponding to negative deviation for goal i with priority j assigned, $d_i^+ \vee 0$ is the positive deviation from the target of goal i, $d_i^- \vee 0$ is the negative deviation from the target of goal i, f_i is a function in goal constraints, g_j is a function in real constraints, b_i is the target value according to goal i, l is the number of priorities, m is the number of goal constraints, and p is the number of real constraints.

7.3 Chance-Constrained Programming

Assume that $\boldsymbol{x}$ is a decision vector, $\boldsymbol{\xi}$ is an uncertain vector, $f(\boldsymbol{x}, \boldsymbol{\xi})$ is a return function, and $g_j(\boldsymbol{x}, \boldsymbol{\xi})$ are constraint functions, $j = 1, 2, \cdots, p$. Since the uncertain constraints $g_j(\boldsymbol{x}, \boldsymbol{\xi}) \le 0, j = 1, 2, \cdots, p$ do not define a deterministic feasible set, it is naturally desired that the uncertain constraints hold with chance α, where α is a specified confidence level. Then we have a chance constraint as follows,

$$
\mathcal{M}\{g_j(\boldsymbol{x}, \boldsymbol{\xi}) \le 0, j = 1, 2, \cdots, p\} \ge \alpha. \tag{7.21}
$$

Maximax Chance-Constrained Programming

If we want to maximize the optimistic value to the uncertain return subject to some chance constraints, we have the following uncertain maximax CCP,

$$
\begin{cases}
\max\limits_{\boldsymbol{x}} \max\limits_{\overline{f}} \overline{f} \\
\text{subject to:} \\
\qquad \mathcal{M}\left\{f(\boldsymbol{x}, \boldsymbol{\xi}) \ge \overline{f}\right\} \ge \beta \\
\qquad \mathcal{M}\left\{g_j(\boldsymbol{x}, \boldsymbol{\xi}) \le 0\right\} \ge \alpha_j, \quad j = 1, 2, \cdots, p
\end{cases}
\tag{7.22}
$$

where α_j and β are specified confidence levels for $j = 1, 2, \cdots, p$, and $\max \overline{f}$ is the β-optimistic return.

In practice, we may have multiple objectives. We thus have the following uncertain maximax chance-constrained multiobjective programming (CCMOP),

$$
\begin{cases}
\max\limits_{\boldsymbol{x}} \left[\max\limits_{\overline{f}_1} \overline{f}_1, \max\limits_{\overline{f}_2} \overline{f}_2, \cdots, \max\limits_{\overline{f}_m} \overline{f}_m \right] \\
\text{subject to:} \\
\qquad \mathcal{M}\left\{ f_i(\boldsymbol{x}, \boldsymbol{\xi}) \ge \overline{f}_i \right\} \ge \beta_i, \quad i = 1, 2, \cdots, m \\
\qquad \mathcal{M}\left\{ g_j(\boldsymbol{x}, \boldsymbol{\xi}) \le 0 \right\} \ge \alpha_j, \quad j = 1, 2, \cdots, p
\end{cases}
\tag{7.23}
$$

where β_i are predetermined confidence levels for $i = 1, 2, \cdots, m$, and $\max \overline{f}_i$ are the β-optimistic values to the return functions $f_i(\boldsymbol{x}, \boldsymbol{\xi})$, $i = 1, 2, \cdots, m$, respectively.

If the priority structure and target levels are set by the decision-maker, then we have a minimin chance-constrained goal programming (CCGP),

$$
\begin{cases}
\min\limits_{\boldsymbol{x}} \sum\limits_{j=1}^{l} P_j \sum\limits_{i=1}^{m} \left(u_{ij} \left(\min\limits_{d_i^+} d_i^+ \vee 0 \right) + v_{ij} \left(\min\limits_{d_i^-} d_i^- \vee 0 \right) \right) \\
\text{subject to:} \\
\qquad \mathcal{M}\left\{ f_i(\boldsymbol{x}, \boldsymbol{\xi}) - b_i \le d_i^+ \right\} \ge \beta_i^+, \quad i = 1, 2, \cdots, m \\
\qquad \mathcal{M}\left\{ b_i - f_i(\boldsymbol{x}, \boldsymbol{\xi}) \le d_i^- \right\} \ge \beta_i^-, \quad i = 1, 2, \cdots, m \\
\qquad \mathcal{M}\left\{ g_j(\boldsymbol{x}, \boldsymbol{\xi}) \le 0 \right\} \ge \alpha_j, \qquad\quad j = 1, 2, \cdots, p
\end{cases}
\tag{7.24}
$$

where P_j is the preemptive priority factor which expresses the relative importance of various goals, $P_j \gg P_{j+1}$, for all j, u_{ij} is the weighting factor corresponding to positive deviation for goal i with priority j assigned, v_{ij} is the weighting factor corresponding to negative deviation for goal i with priority j assigned, $\min d_i^+ \vee 0$ is the β_i^+-optimistic positive deviation from the target of goal i, $\min d_i^- \vee 0$ is the β_i^--optimistic negative deviation from the target of goal i, b_i is the target value according to goal i, and l is the number of priorities.

Minimax Chance-Constrained Programming

If we want to maximize the pessimistic value subject to some chance constraints, we have the following uncertain minimax CCP,

$$
\begin{cases}
\max\limits_{\boldsymbol{x}} \min\limits_{\overline{f}} \overline{f} \\
\text{subject to:} \\
\qquad \mathcal{M}\left\{ f(\boldsymbol{x}, \boldsymbol{\xi}) \le \overline{f} \right\} \ge \beta \\
\qquad \mathcal{M}\left\{ g_j(\boldsymbol{x}, \boldsymbol{\xi}) \le 0 \right\} \ge \alpha_j, \quad j = 1, 2, \cdots, p
\end{cases}
\tag{7.25}
$$

where α_j and β are specified confidence levels for $j = 1, 2, \cdots, p$, and $\min \overline{f}$ is the β-pessimistic return.

If there are multiple objectives, then we have the following uncertain minimax CCMOP,

$$\left\{\begin{array}{l}\max\limits_{\boldsymbol{x}} \left[\min\limits_{\overline{f}_1} \overline{f}_1,\ \min\limits_{\overline{f}_2} \overline{f}_2,\ \cdots,\ \min\limits_{\overline{f}_m} \overline{f}_m\right] \\ \text{subject to:} \\ \qquad \mathcal{M}\left\{f_i(\boldsymbol{x},\boldsymbol{\xi}) \le \overline{f}_i\right\} \ge \beta_i, \quad i=1,2,\cdots,m \\ \qquad \mathcal{M}\left\{g_j(\boldsymbol{x},\boldsymbol{\xi}) \le 0\right\} \ge \alpha_j, \quad j=1,2,\cdots,p\end{array}\right. \tag{7.26}$$

where $\min \overline{f}_i$ are the β_i-pessimistic values to the return functions $f_i(\boldsymbol{x},\boldsymbol{\xi})$, $i=1,2,\cdots,m$, respectively.

We can also formulate a uncertain decision system as an uncertain minimax CCGP according to the priority structure and target levels set by the decision-maker:

$$\left\{\begin{array}{l}\min\limits_{\boldsymbol{x}} \sum\limits_{j=1}^{l} P_j \sum\limits_{i=1}^{m} \left[u_{ij}\left(\max\limits_{d_i^+} d_i^+ \vee 0\right) + v_{ij}\left(\max\limits_{d_i^-} d_i^- \vee 0\right)\right] \\ \text{subject to:} \\ \qquad \mathcal{M}\left\{f_i(\boldsymbol{x},\boldsymbol{\xi}) - b_i \ge d_i^+\right\} \ge \beta_i^+, \quad i=1,2,\cdots,m \\ \qquad \mathcal{M}\left\{b_i - f_i(\boldsymbol{x},\boldsymbol{\xi}) \ge d_i^-\right\} \ge \beta_i^-, \quad i=1,2,\cdots,m \\ \qquad \mathcal{M}\left\{g_j(\boldsymbol{x},\boldsymbol{\xi}) \le 0\right\} \ge \alpha_j, \qquad\quad j=1,2,\cdots,p\end{array}\right. \tag{7.27}$$

where P_j is the preemptive priority factor which expresses the relative importance of various goals, $P_j \gg P_{j+1}$, for all j, u_{ij} is the weighting factor corresponding to positive deviation for goal i with priority j assigned, v_{ij} is the weighting factor corresponding to negative deviation for goal i with priority j assigned, $\max d_i^+ \vee 0$ is the β_i^+-pessimistic positive deviation from the target of goal i, $\max d_i^- \vee 0$ is the β_i^--pessimistic negative deviation from the target of goal i, b_i is the target value according to goal i, and l is the number of priorities.

7.4 Dependent-Chance Programming

This section provides uncertain DCP in which the underlying philosophy is based on selecting the decision with maximum chance to meet the event. A generally uncertain DCP has the following form,

$$\left\{\begin{array}{l}\max \mathcal{M}\left\{h_k(\boldsymbol{x},\boldsymbol{\xi}) \le 0, k=1,2,\cdots,q\right\} \\ \text{subject to:} \\ \qquad g_j(\boldsymbol{x},\boldsymbol{\xi}) \le 0, \quad j=1,2,\cdots,p\end{array}\right. \tag{7.28}$$

where $\boldsymbol{x}$ is an n-dimensional decision vector, $\boldsymbol{\xi}$ is a uncertain vector, the event $\mathcal{E}$ is characterized by $h_k(\boldsymbol{x},\boldsymbol{\xi}) \le 0, k=1,2,\cdots,q$, and the uncertain environment is described by the uncertain constraints $g_j(\boldsymbol{x},\boldsymbol{\xi}) \le 0, j=1,2,\cdots,p$.

If there are multiple events in the uncertain environment, then we have the following uncertain dependent-chance multiobjective programming (DCMOP),

$$\begin{cases} \max \begin{bmatrix} \mathcal{M}\{h_{1k}(\boldsymbol{x},\boldsymbol{\xi}) \le 0, k=1,2,\cdots,q_1\} \\ \mathcal{M}\{h_{2k}(\boldsymbol{x},\boldsymbol{\xi}) \le 0, k=1,2,\cdots,q_2\} \\ \cdots \\ \mathcal{M}\{h_{mk}(\boldsymbol{x},\boldsymbol{\xi}) \le 0, k=1,2,\cdots,q_m\} \end{bmatrix} \\ \text{subject to:} \\ \qquad g_j(\boldsymbol{x},\boldsymbol{\xi}) \le 0, \quad j=1,2,\cdots,p \end{cases} \tag{7.29}$$

where the events $\mathcal{E}_i$ are characterized by $h_{ik}(\boldsymbol{x},\boldsymbol{\xi}) \le 0, k=1,2,\cdots,q_i$, $i = 1,2,\cdots,m$, respectively.

Uncertain dependent-chance goal programming (DCGP) is employed to formulate uncertain decision systems according to the priority structure and target levels set by the decision-maker,

$$\begin{cases} \min \sum\limits_{j=1}^{l} P_j \sum\limits_{i=1}^{m} (u_{ij} d_i^+ \vee 0 + v_{ij} d_i^- \vee 0) \\ \text{subject to:} \\ \quad \mathcal{M}\{h_{ik}(\boldsymbol{x},\boldsymbol{\xi}) \le 0, k=1,2,\cdots,q_i\} - b_i = d_i^+, \quad i=1,2,\cdots,m \\ \quad b_i - \mathcal{M}\{h_{ik}(\boldsymbol{x},\boldsymbol{\xi}) \le 0, k=1,2,\cdots,q_i\} = d_i^-, \quad i=1,2,\cdots,m \\ \quad g_j(\boldsymbol{x},\boldsymbol{\xi}) \le 0, \qquad j=1,2,\cdots,p \end{cases}$$

where P_j is the preemptive priority factor which expresses the relative importance of various goals, $P_j \gg P_{j+1}$, for all j, u_{ij} is the weighting factor corresponding to positive deviation for goal i with priority j assigned, v_{ij} is the weighting factor corresponding to negative deviation for goal i with priority j assigned, $d_i^+ \vee 0$ is the positive deviation from the target of goal i, $d_i^- \vee 0$ is the negative deviation from the target of goal i, g_j is a function in system constraints, b_i is the target value according to goal i, l is the number of priorities, m is the number of goal constraints, and p is the number of system constraints.

7.5 Uncertain Dynamic Programming

Stochastic dynamic programming has been studied widely in the literature. In human decision processes such as diagnosis, psychotherapy, and even design, some parameters may be regarded as uncertain variables. In order to model generally uncertain decision processes, this book proposes a general principle of uncertain dynamic programming, including expected value dynamic programming, chance-constrained dynamic programming and dependent-chance dynamic programming.

Expected Value Dynamic Programming

Consider an N-stage decision system in which $(\boldsymbol{a}_1, \boldsymbol{a}_2, \cdots, \boldsymbol{a}_N)$ represents the state vector, $(\boldsymbol{x}_1, \boldsymbol{x}_2, \cdots, \boldsymbol{x}_N)$ the decision vector, $(\boldsymbol{\xi}_1, \boldsymbol{\xi}_2, \cdots, \boldsymbol{\xi}_N)$ the uncertain vector. We also assume that the state transition function is

$$\boldsymbol{a}_{n+1} = T(\boldsymbol{a}_n, \boldsymbol{x}_n, \boldsymbol{\xi}_n), \quad n = 1, 2, \cdots, N-1. \tag{7.30}$$

In order to maximize the expected return over the horizon, we may use the following expected value dynamic programming,

$$\begin{cases} f_N(\boldsymbol{a}) = \max\limits_{E[g_N(\boldsymbol{a},\boldsymbol{x},\boldsymbol{\xi}_N)]\le 0} E[r_N(\boldsymbol{a},\boldsymbol{x},\boldsymbol{\xi}_N)] \\ f_n(\boldsymbol{a}) = \max\limits_{E[g_n(\boldsymbol{a},\boldsymbol{x},\boldsymbol{\xi}_n)]\le 0} E[r_n(\boldsymbol{a},\boldsymbol{x},\boldsymbol{\xi}_n) + f_{n+1}(T(\boldsymbol{a},\boldsymbol{x},\boldsymbol{\xi}_n))] \\ n \le N-1 \end{cases} \tag{7.31}$$

where r_n are the return functions at the nth stages, $n = 1, 2, \cdots, N$, respectively, and E denotes the expected value operator. This type of uncertain (especially stochastic) dynamic programming has been applied to a wide variety of problems, for example, inventory systems.

Chance-Constrained Dynamic Programming

In order to maximize the optimistic return over the horizon, we may use the following chance-constrained dynamic programming,

$$\begin{cases} f_N(\boldsymbol{a}) = \max\limits_{\mathcal{M}\{g_N(\boldsymbol{a},\boldsymbol{x},\boldsymbol{\xi}_N)\le 0\}\ge\alpha} \overline{r}_N(\boldsymbol{a},\boldsymbol{x},\boldsymbol{\xi}_N) \\ f_n(\boldsymbol{a}) = \max\limits_{\mathcal{M}\{g_n(\boldsymbol{a},\boldsymbol{x},\boldsymbol{\xi}_n)\le 0\}\ge\alpha} \{\overline{r}_n(\boldsymbol{a},\boldsymbol{x},\boldsymbol{\xi}_n) + f_{n+1}(T(\boldsymbol{a},\boldsymbol{x},\boldsymbol{\xi}_n))\} \\ n \le N-1 \end{cases} \tag{7.32}$$

where the functions $\overline{r}_n$ are defined by

$$\overline{r}_n(\boldsymbol{a},\boldsymbol{x},\boldsymbol{\xi}_n) = \sup\left\{\overline{r} \;\middle|\; \mathcal{M}\{r_n(\boldsymbol{a},\boldsymbol{x},\boldsymbol{\xi}_n) \ge \overline{r}\} \ge \beta\right\} \tag{7.33}$$

for $n = 1, 2, \cdots, N$. If we want to maximize the pessimistic return over the horizon, then we must define the functions $\overline{r}_n$ as

$$\overline{r}_n(\boldsymbol{a},\boldsymbol{x},\boldsymbol{\xi}_n) = \inf\left\{\overline{r} \;\middle|\; \mathcal{M}\{r_n(\boldsymbol{a},\boldsymbol{x},\boldsymbol{\xi}_n) \le \overline{r}\} \ge \beta\right\} \tag{7.34}$$

for $n = 1, 2, \cdots, N$.

Dependent-Chance Dynamic Programming

In order to maximize the chance over the horizon, we may employ the following dependent-chance dynamic programming,

$$\begin{cases} f_N(\boldsymbol{a}) = \max\limits_{g_N(\boldsymbol{a},\boldsymbol{x},\boldsymbol{\xi}_N)\le 0} \mathcal{M}\{h_N(\boldsymbol{a},\boldsymbol{x},\boldsymbol{\xi}_N) \le 0\} \\ f_n(\boldsymbol{a}) = \max\limits_{g_n(\boldsymbol{a},\boldsymbol{x},\boldsymbol{\xi}_n)\le 0} \{\mathcal{M}\{h_n(\boldsymbol{a},\boldsymbol{x},\boldsymbol{\xi}_n) \le 0\} + f_{n+1}(T(\boldsymbol{a},\boldsymbol{x},\boldsymbol{\xi}_n))\} \\ n \le N-1 \end{cases}$$

where $h_n(\boldsymbol{a}, \boldsymbol{x}, \boldsymbol{\xi}_n) \le 0$ are the events, and $g_n(\boldsymbol{a}, \boldsymbol{x}, \boldsymbol{\xi}_n) \le 0$ are the uncertain environments at the nth stages, $n = 1, 2, \cdots, N$, respectively.

7.6 Uncertain Multilevel Programming

In order to model generally uncertain decentralized decision systems, this book proposes three types of uncertain multilevel programming, including expected value multilevel programming, chance-constrained multilevel programming and dependent-chance multilevel programming, and provides the concept of Stackelberg-Nash equilibrium to uncertain multilevel programming.

Expected Value Multilevel Programming

Assume that in a decentralized two-level decision system there is one leader and m followers. Let $\boldsymbol{x}$ and $\boldsymbol{y}_i$ be the control vectors of the leader and the ith followers, $i = 1, 2, \cdots, m$, respectively. We also assume that the objective functions of the leader and ith followers are $F(\boldsymbol{x}, \boldsymbol{y}_1, \cdots, \boldsymbol{y}_m, \boldsymbol{\xi})$ and $f_i(\boldsymbol{x}, \boldsymbol{y}_1, \cdots, \boldsymbol{y}_m, \boldsymbol{\xi})$, $i = 1, 2, \cdots, m$, respectively, where $\boldsymbol{\xi}$ is an uncertain (random, fuzzy, hybrid) vector.

Let the feasible set of control vector $\boldsymbol{x}$ of the leader be defined by the expected constraint

$$E[G(\boldsymbol{x}, \boldsymbol{\xi})] \le 0 \tag{7.35}$$

where G is a vector-valued function and 0 is a zero vector. Then for each decision $\boldsymbol{x}$ chosen by the leader, the feasibility of control vectors $\boldsymbol{y}_i$ of the ith followers should be dependent on not only $\boldsymbol{x}$ but also $\boldsymbol{y}_1, \cdots, \boldsymbol{y}_{i-1}, \boldsymbol{y}_{i+1}, \cdots, \boldsymbol{y}_m$, and generally represented by the expected constraints,

$$E[g_i(\boldsymbol{x}, \boldsymbol{y}_1, \boldsymbol{y}_2, \cdots, \boldsymbol{y}_m, \boldsymbol{\xi})] \le 0 \tag{7.36}$$

where g_i are vector-valued functions, $i = 1, 2, \cdots, m$, respectively.

Assume that the leader first chooses his control vector $\boldsymbol{x}$, and the followers determine their control array $(\boldsymbol{y}_1, \boldsymbol{y}_2, \cdots, \boldsymbol{y}_m)$ after that. In order to maximize the expected objective of the leader, we have the following expected value bilevel programming,

$$\left\{\begin{array}{l}
\max\limits_{\boldsymbol{x}} E[F(\boldsymbol{x}, \boldsymbol{y}_1^*, \boldsymbol{y}_2^*, \cdots, \boldsymbol{y}_m^*, \boldsymbol{\xi})] \\
\text{subject to:} \\
\qquad E[G(\boldsymbol{x}, \boldsymbol{\xi})] \le 0 \\
\qquad (\boldsymbol{y}_1^*, \boldsymbol{y}_2^*, \cdots, \boldsymbol{y}_m^*) \text{ solves problems } (i = 1, 2, \cdots, m) \\
\qquad \left\{\begin{array}{l}
\max\limits_{\boldsymbol{y}_i} E[f_i(\boldsymbol{x}, \boldsymbol{y}_1, \boldsymbol{y}_2, \cdots, \boldsymbol{y}_m, \boldsymbol{\xi})] \\
\text{subject to:} \\
\qquad E[g_i(\boldsymbol{x}, \boldsymbol{y}_1, \boldsymbol{y}_2, \cdots, \boldsymbol{y}_m, \boldsymbol{\xi})] \le 0.
\end{array}\right.
\end{array}\right. \tag{7.37}$$

Definition 7.9. *Let $\boldsymbol{x}$ be a feasible control vector of the leader. A Nash equilibrium of followers is the feasible array $(\boldsymbol{y}_1^*, \boldsymbol{y}_2^*, \cdots, \boldsymbol{y}_m^*)$ with respect to $\boldsymbol{x}$ if*

$$\begin{aligned} &E[f_i(\boldsymbol{x}, \boldsymbol{y}_1^*, \cdots, \boldsymbol{y}_{i-1}^*, \boldsymbol{y}_i, \boldsymbol{y}_{i+1}^*, \cdots, \boldsymbol{y}_m^*, \boldsymbol{\xi})] \\ &\quad \le E[f_i(\boldsymbol{x}, \boldsymbol{y}_1^*, \cdots, \boldsymbol{y}_{i-1}^*, \boldsymbol{y}_i^*, \boldsymbol{y}_{i+1}^*, \cdots, \boldsymbol{y}_m^*, \boldsymbol{\xi})] \end{aligned} \tag{7.38}$$

for any feasible array $(\boldsymbol{y}_1^*, \cdots, \boldsymbol{y}_{i-1}^*, \boldsymbol{y}_i, \boldsymbol{y}_{i+1}^*, \cdots, \boldsymbol{y}_m^*)$ and $i = 1, 2, \cdots, m$.

Definition 7.10. *Suppose that $\boldsymbol{x}^*$ is a feasible control vector of the leader and $(\boldsymbol{y}_1^*, \boldsymbol{y}_2^*, \cdots, \boldsymbol{y}_m^*)$ is a Nash equilibrium of followers with respect to $\boldsymbol{x}^*$. We call the array $(\boldsymbol{x}^*, \boldsymbol{y}_1^*, \boldsymbol{y}_2^*, \cdots, \boldsymbol{y}_m^*)$ a Stackelberg-Nash equilibrium to the expected value bilevel programming (7.37) if and only if*

$$E[F(\overline{\boldsymbol{x}}, \overline{\boldsymbol{y}}_1, \overline{\boldsymbol{y}}_2, \cdots, \overline{\boldsymbol{y}}_m, \boldsymbol{\xi})] \le E[F(\boldsymbol{x}^*, \boldsymbol{y}_1^*, \boldsymbol{y}_2^*, \cdots, \boldsymbol{y}_m^*, \boldsymbol{\xi})] \tag{7.39}$$

for any feasible control vector $\overline{\boldsymbol{x}}$ and the Nash equilibrium $(\overline{\boldsymbol{y}}_1, \overline{\boldsymbol{y}}_2, \cdots, \overline{\boldsymbol{y}}_m)$ with respect to $\overline{\boldsymbol{x}}$.

Chance-Constrained Multilevel Programming

In order to maximize the optimistic return subject to the chance constraint, we may use the following chance-constrained bilevel programming,

$$\begin{cases} \max\limits_{\boldsymbol{x}} \max\limits_{\overline{F}} \overline{F} \\ \text{subject to:} \\ \qquad \mathcal{M}\{F(\boldsymbol{x}, \boldsymbol{y}_1^*, \boldsymbol{y}_2^*, \cdots, \boldsymbol{y}_m^*, \boldsymbol{\xi}) \ge \overline{F}\} \ge \beta \\ \qquad \mathcal{M}\{G(\boldsymbol{x}, \boldsymbol{\xi}) \le 0\} \ge \alpha \\ \qquad (\boldsymbol{y}_1^*, \boldsymbol{y}_2^*, \cdots, \boldsymbol{y}_m^*) \text{ solves problems } (i = 1, 2, \cdots, m) \\ \qquad \begin{cases} \max\limits_{\boldsymbol{y}_i} \max\limits_{\overline{f}_i} \overline{f}_i \\ \text{subject to:} \\ \qquad \mathcal{M}\{f_i(\boldsymbol{x}, \boldsymbol{y}_1, \boldsymbol{y}_2, \cdots, \boldsymbol{y}_m, \boldsymbol{\xi}) \ge \overline{f}_i\} \ge \beta_i \\ \qquad \mathcal{M}\{g_i(\boldsymbol{x}, \boldsymbol{y}_1, \boldsymbol{y}_2, \cdots, \boldsymbol{y}_m, \boldsymbol{\xi}) \le 0\} \ge \alpha_i \end{cases} \end{cases} \tag{7.40}$$

where $\alpha, \beta, \alpha_i, \beta_i$, $i = 1, 2, \cdots, m$ are predetermined confidence levels.

Definition 7.11. *Let $\boldsymbol{x}$ be a feasible control vector of the leader. A Nash equilibrium of followers is the feasible array $(\boldsymbol{y}_1^*, \boldsymbol{y}_2^*, \cdots, \boldsymbol{y}_m^*)$ with respect to $\boldsymbol{x}$ if*

$$\begin{aligned} &\overline{f}_i(\boldsymbol{x}, \boldsymbol{y}_1^*, \cdots, \boldsymbol{y}_{i-1}^*, \boldsymbol{y}_i, \boldsymbol{y}_{i+1}^*, \cdots, \boldsymbol{y}_m^*) \\ &\quad \le \overline{f}_i(\boldsymbol{x}, \boldsymbol{y}_1^*, \cdots, \boldsymbol{y}_{i-1}^*, \boldsymbol{y}_i^*, \boldsymbol{y}_{i+1}^*, \cdots, \boldsymbol{y}_m^*) \end{aligned} \tag{7.41}$$

for any feasible array $(\boldsymbol{y}_1^*, \cdots, \boldsymbol{y}_{i-1}^*, \boldsymbol{y}_i, \boldsymbol{y}_{i+1}^*, \cdots, \boldsymbol{y}_m^*)$ and $i = 1, 2, \cdots, m$.

Definition 7.12. *Suppose that $\boldsymbol{x}^*$ is a feasible control vector of the leader and $(\boldsymbol{y}_1^*, \boldsymbol{y}_2^*, \cdots, \boldsymbol{y}_m^*)$ is a Nash equilibrium of followers with respect to $\boldsymbol{x}^*$.*

The array $(\boldsymbol{x}^*, \boldsymbol{y}_1^*, \boldsymbol{y}_2^*, \cdots, \boldsymbol{y}_m^*)$ is called a Stackelberg-Nash equilibrium to the chance-constrained bilevel programming (7.40) if and only if,

$$\overline{F}(\overline{\boldsymbol{x}}, \overline{\boldsymbol{y}}_1, \overline{\boldsymbol{y}}_2, \cdots, \overline{\boldsymbol{y}}_m) \le \overline{F}(\boldsymbol{x}^*, \boldsymbol{y}_1^*, \boldsymbol{y}_2^*, \cdots, \boldsymbol{y}_m^*) \tag{7.42}$$

for any feasible control vector $\overline{\boldsymbol{x}}$ and the Nash equilibrium $(\overline{\boldsymbol{y}}_1, \overline{\boldsymbol{y}}_2, \cdots, \overline{\boldsymbol{y}}_m)$ with respect to $\overline{\boldsymbol{x}}$.

In order to maximize the pessimistic return, we have the following minimax chance-constrained bilevel programming,

$$\begin{cases} \max\limits_{\boldsymbol{x}} \min\limits_{\overline{F}} \overline{F} \\ \text{subject to:} \\ \qquad \mathcal{M}\{F(\boldsymbol{x}, \boldsymbol{y}_1^*, \boldsymbol{y}_2^*, \cdots, \boldsymbol{y}_m^*, \boldsymbol{\xi}) \le \overline{F}\} \ge \beta \\ \qquad \mathcal{M}\{G(\boldsymbol{x}, \boldsymbol{\xi}) \le 0\} \ge \alpha \\ \qquad (\boldsymbol{y}_1^*, \boldsymbol{y}_2^*, \cdots, \boldsymbol{y}_m^*) \text{ solves problems } (i = 1, 2, \cdots, m) \\ \qquad \begin{cases} \max\limits_{\boldsymbol{y}_i} \min\limits_{\overline{f}_i} \overline{f}_i \\ \text{subject to:} \\ \qquad \mathcal{M}\{f_i(\boldsymbol{x}, \boldsymbol{y}_1, \boldsymbol{y}_2, \cdots, \boldsymbol{y}_m, \boldsymbol{\xi}) \le \overline{f}_i\} \ge \beta_i \\ \qquad \mathcal{M}\{g_i(\boldsymbol{x}, \boldsymbol{y}_1, \boldsymbol{y}_2, \cdots, \boldsymbol{y}_m, \boldsymbol{\xi}) \le 0\} \ge \alpha_i. \end{cases} \end{cases} \tag{7.43}$$

Dependent-Chance Multilevel Programming

Let $H(\boldsymbol{x}, \boldsymbol{y}_1, \boldsymbol{y}_2, \cdots, \boldsymbol{y}_m, \boldsymbol{\xi}) \le 0$ and $h_i(\boldsymbol{x}, \boldsymbol{y}_1, \boldsymbol{y}_2, \cdots, \boldsymbol{y}_m, \boldsymbol{\xi}) \le 0$ be the tasks of the leader and ith followers, $i = 1, 2, \cdots, m$, respectively. In order to maximize the chance functions of the leader and followers, we have the following dependent-chance bilevel programming,

$$\begin{cases} \max\limits_{\boldsymbol{x}} \mathcal{M}\{H(\boldsymbol{x}, \boldsymbol{y}_1^*, \boldsymbol{y}_2^*, \cdots, \boldsymbol{y}_m^*, \boldsymbol{\xi}) \le 0\} \\ \text{subject to:} \\ \qquad G(\boldsymbol{x}, \boldsymbol{\xi}) \le 0 \\ \qquad (\boldsymbol{y}_1^*, \boldsymbol{y}_2^*, \cdots, \boldsymbol{y}_m^*) \text{ solves problems } (i = 1, 2, \cdots, m) \\ \qquad \begin{cases} \max\limits_{\boldsymbol{y}_i} \mathcal{M}\{h_i(\boldsymbol{x}, \boldsymbol{y}_1, \boldsymbol{y}_2, \cdots, \boldsymbol{y}_m, \boldsymbol{\xi}) \le 0\} \\ \text{subject to:} \\ \qquad g_i(\boldsymbol{x}, \boldsymbol{y}_1, \boldsymbol{y}_2, \cdots, \boldsymbol{y}_m, \boldsymbol{\xi}) \le 0. \end{cases} \end{cases} \tag{7.44}$$

Definition 7.13. *Let $\boldsymbol{x}$ be a control vector of the leader. We call the array $(\boldsymbol{y}_1^*, \boldsymbol{y}_2^*, \cdots, \boldsymbol{y}_m^*)$ a Nash equilibrium of followers with respect to $\boldsymbol{x}$ if*

$$\begin{aligned} &\mathcal{M}\{h_i(\boldsymbol{x}, \boldsymbol{y}_1^*, \cdots, \boldsymbol{y}_{i-1}^*, \boldsymbol{y}_i, \boldsymbol{y}_{i+1}^*, \cdots, \boldsymbol{y}_m^*, \boldsymbol{\xi}) \le 0\} \\ &\quad \le \mathcal{M}\{h_i(\boldsymbol{x}, \boldsymbol{y}_1^*, \cdots, \boldsymbol{y}_{i-1}^*, \boldsymbol{y}_i^*, \boldsymbol{y}_{i+1}^*, \cdots, \boldsymbol{y}_m^*, \boldsymbol{\xi}) \le 0\} \end{aligned} \tag{7.45}$$

subject to the uncertain environment $g_i(\boldsymbol{x}, \boldsymbol{y}_1, \boldsymbol{y}_2, \cdots, \boldsymbol{y}_m, \boldsymbol{\xi}) \le 0, i = 1, 2, \cdots, m$ *for any array* $(\boldsymbol{y}_1^*, \cdots, \boldsymbol{y}_{i-1}^*, \boldsymbol{y}_i, \boldsymbol{y}_{i+1}^*, \cdots, \boldsymbol{y}_m^*)$ *and* $i = 1, 2, \cdots, m$.

Definition 7.14. *Let* $\boldsymbol{x}^*$ *be a control vector of the leader, and* $(\boldsymbol{y}_1^*, \boldsymbol{y}_2^*, \cdots, \boldsymbol{y}_m^*)$ *a Nash equilibrium of followers with respect to* $\boldsymbol{x}^*$. *Then* $(\boldsymbol{x}^*, \boldsymbol{y}_1^*, \boldsymbol{y}_2^*, \cdots, \boldsymbol{y}_m^*)$ *is called a Stackelberg-Nash equilibrium to the dependent-chance bilevel programming (7.44) if and only if,*

$$\mathcal{M}\{H(\overline{\boldsymbol{x}}, \overline{\boldsymbol{y}}_1, \overline{\boldsymbol{y}}_2, \cdots, \overline{\boldsymbol{y}}_m, \boldsymbol{\xi}) \le 0\} \le \mathcal{M}\{H(\boldsymbol{x}^*, \boldsymbol{y}_1^*, \boldsymbol{y}_2^*, \cdots, \boldsymbol{y}_m^*, \boldsymbol{\xi}) \le 0\}$$

subject to the uncertain environment $G(\boldsymbol{x}, \boldsymbol{\xi}) \le 0$ *for any control vector* $\overline{\boldsymbol{x}}$ *and the Nash equilibrium* $(\overline{\boldsymbol{y}}_1, \overline{\boldsymbol{y}}_2, \cdots, \overline{\boldsymbol{y}}_m)$ *with respect to* $\overline{\boldsymbol{x}}$.

7.7 Ψ Graph of Uncertain Programming

There are many possible ways that we can classify uncertain programming models. For example, we may classify them according to the state of knowledge about information, modeling structure, and uncertainty-handling philosophy. Let us now briefly outline some important aspects to be considered when discussing each of these.

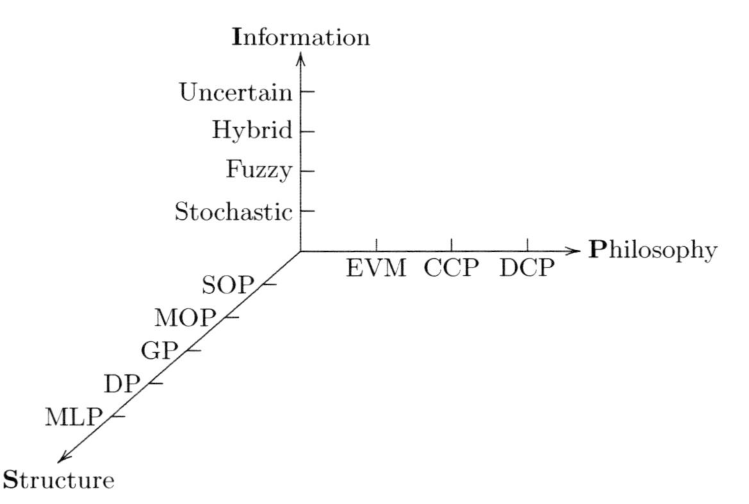

Fig. 7.1 Ψ Graph for Uncertain Programming Classifications. Any type of uncertain programming may be represented by the Ψ graph which is essentially a coordinate system (**P**hilosophy, **S**tructure, **I**nformation). For example, the plane "**P**=CCP" represents chance-constrained programming; the plane "**I**=Stochastic" represents stochastic programming; the point "(**P**,**S**,**I**)=(DCP, GP, Fuzzy)" represents fuzzy dependent-chance goal programming.

1. State of knowledge about information
    a. Stochastic variable
    b. Fuzzy variable
    c. Hybrid variable
    d. Uncertain variable
2. Modeling structure
    a. Single-objective programming
    b. Multiobjective programming
    c. Goal programming
    d. Dynamic programming
    e. Multilevel programming
3. Uncertainty-handling philosophy
    a. Expected value model
    b. Chance-constrained programming
    c. Dependent-chance programming

Chapter 8
System Reliability Design

One of the approaches to improve system reliability is to provide redundancy for components in a system. There are two ways to provide component redundancy: parallel redundancy and standby redundancy. In parallel redundancy, all redundant elements are required to operate simultaneously. This method is usually used when element replacements are not permitted during the system operation. In standby redundancy, one of the redundant elements begins to work only when the active element fails. This method is usually employed when the replacement is allowable and can be finished immediately.

The system reliability design problem is to determine the optimal number of redundant elements for each component so as to optimize some system performance.

8.1 Problem Description

Assume that a system consists of n components, and the ith components consist of x_i redundant elements, $i = 1, 2, \cdots, n$, respectively. For example, Figure 8.1 shows a bridge system in which we suppose that redundant elements are in standby for the first and second components, and are in parallel for the third to fifth components.

The first problem is how to estimate the system lifetime when the value of the vector $\boldsymbol{x} = (x_1, x_2, \cdots, x_n)$ is determined. For such a given decision vector $\boldsymbol{x}$, suppose that the redundant elements j operating in components i have lifetimes ξ_{ij}, $j = 1, 2, \cdots, x_i$, $i = 1, 2, \cdots, n$, respectively. For convenience, we use the vector

$$\boldsymbol{\xi} = (\xi_{11}, \xi_{12}, \cdots, \xi_{1x_1}, \xi_{21}, \xi_{22}, \cdots, \xi_{2x_2}, \cdots, \xi_{n1}, \xi_{n2}, \cdots, \xi_{nx_n})$$

to denote the lifetimes of all redundant elements in the system. For parallel redundancy components, the lifetimes are

$$T_i(\boldsymbol{x}, \boldsymbol{\xi}) = \max_{1 \le j \le x_i} \xi_{ij}, \quad i = 1, 2, \cdots, n.$$

B. Liu: Theory and Practice of Uncertain Programming, STUDFUZZ 239, pp. 129–137.
springerlink.com

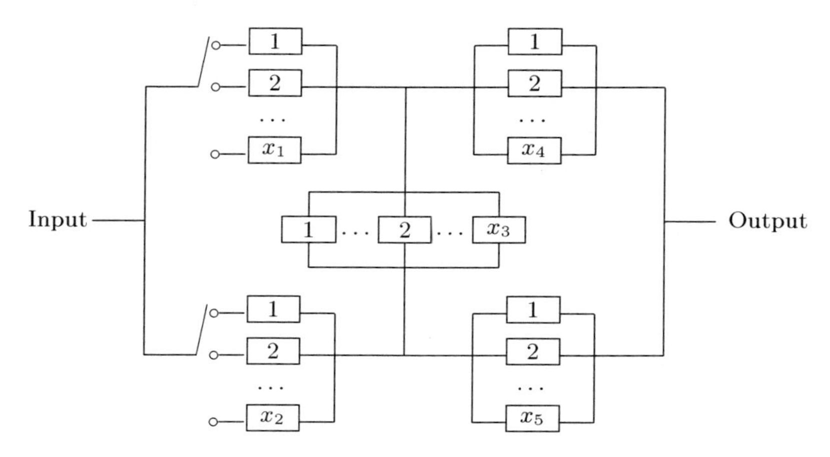

Fig. 8.1 A Bridge System

For standby redundancy components, the lifetimes are

$$T_i(\boldsymbol{x}, \boldsymbol{\xi}) = \sum_{j=1}^{x_i} \xi_{ij}, \quad i = 1, 2, \cdots, n.$$

How do we calculate the system lifetime $T(\boldsymbol{x}, \boldsymbol{\xi})$? It is problem-dependent. For the bridge system shown in Figure 8.1, since the system works if and only if there is a path of working components from the input of the system to the output, the system lifetime is

$$T(\boldsymbol{x}, \boldsymbol{\xi}) = \max \left\{ \begin{array}{c} T_1(\boldsymbol{x}, \boldsymbol{\xi}) \wedge T_4(\boldsymbol{x}, \boldsymbol{\xi}) \\ T_2(\boldsymbol{x}, \boldsymbol{\xi}) \wedge T_5(\boldsymbol{x}, \boldsymbol{\xi}) \\ T_2(\boldsymbol{x}, \boldsymbol{\xi}) \wedge T_3(\boldsymbol{x}, \boldsymbol{\xi}) \wedge T_4(\boldsymbol{x}, \boldsymbol{\xi}) \\ T_1(\boldsymbol{x}, \boldsymbol{\xi}) \wedge T_3(\boldsymbol{x}, \boldsymbol{\xi}) \wedge T_5(\boldsymbol{x}, \boldsymbol{\xi}) \end{array} \right\}.$$

8.2 Stochastic Models

In practice, the lifetime $\boldsymbol{\xi}$ of elements is usually a random vector. Thus the component lifetimes $T_i(\boldsymbol{x}, \boldsymbol{\xi})$ and system lifetime $T(\boldsymbol{x}, \boldsymbol{\xi})$ are also random variables for $i = 1, 2, \cdots, n$.

Stochastic Expected Lifetime Maximization Model

One of system performances is the *expected lifetime* $E[T(\boldsymbol{x}, \boldsymbol{\xi})]$. It is obvious that the greater the expected lifetime $E[T(\boldsymbol{x}, \boldsymbol{\xi})]$, the better the decision $\boldsymbol{x}$.

Let us consider the bridge system shown in Figure 8.1. For simplicity, we suppose that there is only one type of element to be selected for each

Table 8.1 Random Lifetimes and Prices of Elements

Type	1	2	3	4	5
Lifetime	$\mathcal{EXP}(20)$	$\mathcal{EXP}(30)$	$\mathcal{EXP}(40)$	$\mathcal{EXP}(50)$	$\mathcal{EXP}(60)$
Price	50	60	70	80	90

component. The lifetimes of elements are assumed to be exponentially distributed random variables $\mathcal{EXP}(\beta)$ shown in Table 8.1. The decision vector may be represented by $\boldsymbol{x} = (x_1, x_2, \cdots, x_5)$, where x_i denote the numbers of the i-th types of elements selected, $i = 1, 2, \cdots, 5$, respectively.

Another important problem is to compute the cost spent for the system. It follows from Table 8.1 that the the total cost

$$C(\boldsymbol{x}) = 50x_1 + 60x_2 + 70x_3 + 80x_4 + 90x_5.$$

If the total capital available is 600, then we have a constraint $C(\boldsymbol{x}) \le 600$.

For the redundancy system, since we wish to maximize the expected lifetime $E[T(\boldsymbol{x}, \boldsymbol{\xi})]$ subject to the cost constraint, we have the following stochastic expected lifetime maximization model,

$$\begin{cases} \max E[T(\boldsymbol{x}, \boldsymbol{\xi})] \\ \text{subject to:} \\ \qquad C(\boldsymbol{x}) \le 600 \\ \qquad \boldsymbol{x} \ge 1, \ \text{integer vector.} \end{cases} \tag{8.1}$$

Hybrid Intelligent Algorithm

In order to solve this type of model, we may employ the hybrid intelligent algorithm documented in Chapter 4 provided that the initialization, crossover and mutation operations are revised as follows.

Generally speaking, we use an integer vector $V = (x_1, x_2, \cdots, x_n)$ as a chromosome to represent a solution $\boldsymbol{x}$, where x_i are positive integers, $i = 1, 2, \cdots, n$. First we set all genes x_i as 1, $i = 1, 2, \cdots, n$, and form a chromosome V. Then we randomly sample an integer i between 1 and n, and the gene x_i of V is replaced with $x_i + 1$. We repeat this process until the chromosome V is proven to be infeasible. We take the last feasible chromosome as an initial chromosome.

We do the crossover operation on V_1 and V_2 in the following way. Write

$$V_1 = \left(x_1^{(1)}, x_2^{(1)}, \cdots, x_n^{(1)}\right), \quad V_2 = \left(x_1^{(2)}, x_2^{(2)}, \cdots, x_n^{(2)}\right)$$

and randomly generate two integers between 1 and n as the crossover points denoted by n_1 and n_2 such that $n_1 < n_2$. Then we exchange the genes of the chromosomes V_1 and V_2 between n_1 and n_2 and produce two children as follows,

$$V_1' = \left(x_1^{(1)}, x_2^{(1)}, \cdots, x_{n_1-1}^{(1)}, x_{n_1}^{(2)}, \cdots, x_{n_2}^{(2)}, x_{n_2+1}^{(1)}, \cdots, x_n^{(1)}\right),$$

$$V_2' = \left(x_1^{(2)}, x_2^{(2)}, \cdots, x_{n_1-1}^{(2)}, x_{n_1}^{(1)}, \cdots, x_{n_2}^{(1)}, x_{n_2+1}^{(2)}, \cdots, x_n^{(2)}\right).$$

If the child V_1' is infeasible, then we use the following strategy to repair it and make it feasible. At first, we randomly sample an integer i between 1 and n, and then replace the gene x_i of V_1' with $x_i - 1$ provided that $x_i \geq 2$. Repeat this process until the revised chromosome V_1' is feasible. If the child V_1' is proven to be feasible, then we revise it in the following way. We randomly sample an integer i between 1 and n, and the gene x_i of V_1' will be replaced with $x_i + 1$. We repeat this process until the revised chromosome is infeasible, and take the last feasible chromosome as V_1'. A similar revising process will be made on V_2'.

For each selected parent $V = (x_1, x_2, \cdots, x_n)$, we mutate it by the following way. We randomly choose two mutation positions n_1 and n_2 between 1 and n such that $n_1 < n_2$, then we set all genes x_j of V as 1 for $j = n_1, n_1 + 1, \cdots, n_2$, and form a new one

$$V' = (x_1, \cdots, x_{n_1-1}, 1, \cdots, 1, x_{n_2+1}, \cdots, x_n).$$

We will modify V' by the following process. We randomly sample an integer i between n_1 and n_2, and the gene x_i of V' will be replaced with $x_i + 1$. We repeat this process until the revised chromosome is infeasible. Finally, we replace the parent V' with the last feasible chromosome.

Optimal Solution of Model (8.1)

In order to solve the stochastic expected lifetime maximization model (8.1), we deal with the uncertain function

$$U : \boldsymbol{x} \to E[T(\boldsymbol{x}, \boldsymbol{\xi})]$$

by stochastic simulation. Then, we embed the stochastic simulation into a GA and produce a hybrid intelligent algorithm. A run of the hybrid intelligent algorithm (5000 cycles in simulation and 1000 generations in GA) shows that the optimal solution is

$$\boldsymbol{x}^* = (1, 3, 1, 1, 2)$$

whose expected system lifetime is $E[T(\boldsymbol{x}^*, \boldsymbol{\xi})] = 62.5$, and total cost is $C(\boldsymbol{x}^*) = 560$.

Stochastic α-Lifetime Maximization Model

The second type of system performance is the *α-lifetime* defined as the largest value $\overline{T}$ satisfying $\Pr\left\{T(\boldsymbol{x},\boldsymbol{\xi}) \ge \overline{T}\right\} \ge \alpha$, where α is a predetermined confidence level.

This section will model redundancy optimization under this criterion. Consider the bridge system shown in Figure 8.1. The aim is to determine the optimal numbers of the redundant elements so as to maximize the α-lifetime under the cost constraint. Zhao and Liu [330] presented the following stochastic α-lifetime maximization model,

$$
\begin{cases}
\max \overline{T} \\
\text{subject to:} \\
\qquad \Pr\left\{T(\boldsymbol{x},\boldsymbol{\xi}) \ge \overline{T}\right\} \ge \alpha \\
\qquad C(\boldsymbol{x}) \le 600 \\
\qquad \boldsymbol{x} \ge 1, \text{ integer vector.}
\end{cases}
\tag{8.2}
$$

For each observational vector $\boldsymbol{\xi}$ of lifetimes of elements, we may estimate the system lifetime $T(\boldsymbol{x},\boldsymbol{\xi})$. We use the stochastic simulation to deal with the uncertain function

$$
U : \boldsymbol{x} \to \max\left\{\overline{T} \mid \Pr\left\{T(\boldsymbol{x},\boldsymbol{\xi}) \ge \overline{T}\right\} \ge \alpha\right\}.
$$

Then the stochastic simulation is embedded into a GA to form a hybrid intelligent algorithm.

When $\alpha = 0.9$, a run of the hybrid intelligent algorithm (5000 cycles in simulation and 1000 generations in GA) shows that the optimal solution is

$$
\boldsymbol{x}^* = (3, 2, 1, 2, 1)
$$

with 0.9-system lifetime $\overline{T}^* = 25.7$, and the total cost $C(\boldsymbol{x}^*) = 590$.

Stochastic System Reliability Model

The third type of system performance is the *system reliability* $\Pr\{T(\boldsymbol{x},\boldsymbol{\xi}) \ge T^0\}$, which is the probability that the system lifetime is greater than or equal to the given time T^0.

If one wants to maximize the system reliability under a cost constraint, then use the following stochastic system reliability model,

$$
\begin{cases}
\max \Pr\{T(\boldsymbol{x},\boldsymbol{\xi}) \ge T^0\} \\
\text{subject to:} \\
\qquad C(\boldsymbol{x}) \le 600 \\
\qquad \boldsymbol{x} \ge 1, \text{ integer vector.}
\end{cases}
\tag{8.3}
$$

In order to solve this model, we have to deal with the uncertain function

$$U: \boldsymbol{x} \to \Pr\{T(\boldsymbol{x}, \boldsymbol{\xi}) \geq T^0\}$$

by stochastic simulation. Then we embed the stochastic simulation into a GA to produce a hybrid intelligent algorithm.

A run of the hybrid intelligent algorithm (15000 cycles in simulation and 300 generations in GA) shows that the optimal solution is

$$\boldsymbol{x}^* = (4, 1, 1, 2, 1)$$

with $\Pr\{T(\boldsymbol{x}^*, \boldsymbol{\xi}) \geq T^0\} = 0.85$, and the total cost $C(x^*) = 580$.

8.3 Fuzzy Models

Although stochastic programming has been successfully applied in redundancy optimization, many problems require subjective judgment either due to the lack of data or due to the extreme complexity of the system. This fact motives us to apply fuzzy programming to redundancy optimization problems in which the lifetimes of elements are treated as fuzzy variables.

Fuzzy Expected Lifetime Maximization Model

Let us reconsider the bridge system shown in Figure 8.1. The lifetimes of elements are assumed to be triangular fuzzy variables shown in Table 8.2. The decision vector is represented by $\boldsymbol{x} = (x_1, x_2, \cdots, x_5)$, where x_i denote the numbers of the i-th types of elements selected, $i = 1, 2, \cdots, 5$, respectively. It follows from Table 8.2 that the the total cost

$$C(\boldsymbol{x}) = 50x_1 + 60x_2 + 70x_3 + 80x_4 + 90x_5.$$

If the total capital available is 600, then we have a constraint $C(\boldsymbol{x}) \leq 600$.

For such a standby redundancy system, we define *expected lifetime* as $E[T(\boldsymbol{x}, \boldsymbol{\xi})]$. If we wish to maximize the expected lifetime $E[T(\boldsymbol{x}, \boldsymbol{\xi})]$, then we have the following fuzzy expected lifetime maximization model (Zhao and Liu [333]),

Table 8.2 Fuzzy Lifetimes and Prices of Elements

Type	1	2	3	4	5
Lifetime	$(10, 20, 40)$	$(20, 30, 50)$	$(30, 40, 60)$	$(40, 50, 60)$	$(50, 60, 80)$
Price	50	60	70	80	90

$$
\begin{cases}
\max E[T(\boldsymbol{x}, \boldsymbol{\xi})] \\
\text{subject to:} \\
\qquad C(\boldsymbol{x}) \leq 600 \\
\qquad \boldsymbol{x} \geq 1, \text{ integer vector.}
\end{cases}
\tag{8.4}
$$

A run of the hybrid intelligent algorithm (15000 cycles in simulation, 300 generations in GA) shows that the optimal solution is

$$\boldsymbol{x}^* = (2, 4, 1, 1, 1)$$

whose expected system lifetime is $E[T(\boldsymbol{x}^*, \boldsymbol{\xi})] = 60.7$, and the total cost is $C(x^*) = 580$.

Fuzzy α-Lifetime Maximization Model

By *α-lifetime* we mean the largest value $\overline{T}$ satisfying $\text{Cr}\left\{T(\boldsymbol{x}, \boldsymbol{\xi}) \geq \overline{T}\right\} \geq \alpha$, where α is a predetermined confidence level.

If the decision maker wants to maximize the α-lifetime subject to the cost constraint, then we have the following fuzzy α-lifetime maximization model (Zhao and Liu [333]),

$$
\begin{cases}
\max \overline{T} \\
\text{subject to:} \\
\qquad \text{Cr}\left\{T(\boldsymbol{x}, \boldsymbol{\xi}) \geq \overline{T}\right\} \geq \alpha \\
\qquad C(\boldsymbol{x}) \leq 600 \\
\qquad \boldsymbol{x} \geq 1, \text{ integer vector.}
\end{cases}
\tag{8.5}
$$

When $\alpha = 0.8$, a run of the hybrid intelligent algorithm (15000 cycles in simulation and 300 generations in GA) shows that the optimal solution is

$$\boldsymbol{x}^* = (2, 4, 1, 1, 1)$$

whose 0.8-system lifetime $\overline{T}^* = 53.1$, and the total cost $C(\boldsymbol{x}^*) = 580$.

Fuzzy System Reliability Model

By *system reliability* we mean $\text{Cr}\{T(\boldsymbol{x}, \boldsymbol{\xi}) \geq T^0\}$, i.e., the credibility that the system lifetime is greater than or equal to the given time T^0.

If one wants to maximize the system reliability under a cost constraint, then use the following fuzzy system reliability model (Zhao and Liu [333]),

$$
\begin{cases}
\max \text{Cr}\{T(\boldsymbol{x}, \boldsymbol{\xi}) \geq T^0\} \\
\text{subject to:} \\
\qquad C(\boldsymbol{x}) \leq 600 \\
\qquad \boldsymbol{x} \geq 1, \text{ integer vector.}
\end{cases}
\tag{8.6}
$$

When $T^0 = 50$, a run of the hybrid intelligent algorithm (15000 cycles in simulation and 300 generations in GA) shows that the optimal solution is

$$\boldsymbol{x}^* = (2, 4, 1, 1, 1)$$

with $\text{Cr}\{T(\boldsymbol{x}^*, \boldsymbol{\xi}) \geq T^0\} = 0.95$ and the total cost $C(\boldsymbol{x}^*) = 580$.

8.4 Hybrid Models

In a classical system reliability design problem, the element lifetimes are assumed to be random variables or fuzzy variables. Although this assumption has been accepted and accorded with the facts in widespread cases, it is not appropriate in a vast range of situations. In many practical situations, the fuzziness and randomness of the element lifetimes are often mixed up with each other. For example, the element lifetimes are assumed to be exponentially distributed random variables with fuzzy parameters. In this case, fuzziness and randomness of the element lifetimes are required to be considered simultaneously and the effectiveness of the classical redundancy optimization theory is lost.

Hybrid Expected Lifetime Maximization Model

Now we suppose that the element lifetimes are hybrid variables. For this case, the system lifetime $T(\boldsymbol{x}, \boldsymbol{\xi})$ is also a hybrid variable. If we wish to maximize the expected lifetime $E[T(\boldsymbol{x}, \boldsymbol{\xi})]$, then we have the following hybrid expected lifetime maximization model,

$$\begin{cases} \max E[T(\boldsymbol{x}, \boldsymbol{\xi})] \\ \text{subject to:} \\ \qquad C(\boldsymbol{x}) \leq \overline{C} \\ \qquad \boldsymbol{x} \geq 1, \text{ integer vector} \end{cases} \tag{8.7}$$

where $\overline{C}$ is the total capital.

Hybrid α-Lifetime Maximization Model

We define the largest value $\overline{T}$ satisfying $\text{Ch}\left\{T(\boldsymbol{x}, \boldsymbol{\xi}) \geq \overline{T}\right\} \geq \alpha$ as the α-*lifetime* of the system, where α is a predetermined confidence level. If the decision maker wants to maximize the α-lifetime subject to the cost constraint, then we have the following hybrid α-lifetime maximization model,

$$\begin{cases} \max \overline{T} \\ \text{subject to:} \\ \qquad \text{Ch}\left\{T(\boldsymbol{x}, \boldsymbol{\xi}) \geq \overline{T}\right\} \geq \alpha \\ \qquad C(\boldsymbol{x}) \leq \overline{C} \\ \qquad \boldsymbol{x} \geq 1, \text{ integer vector} \end{cases} \tag{8.8}$$

where $\overline{C}$ is the total capital.

Hybrid System Reliability Model

By *system reliability* we mean $\text{Ch}\{T(\boldsymbol{x},\boldsymbol{\xi}) \ge T^0\}$, i.e., the chance that the system lifetime is greater than or equal to the given time T^0. If one wants to maximize the system reliability under a cost constraint, then use the following hybrid system reliability model:

$$\begin{cases} \max \text{Ch}\{T(\boldsymbol{x},\boldsymbol{\xi}) \ge T^0\} \\ \text{subject to:} \\ \qquad C(\boldsymbol{x}) \le \overline{C} \\ \qquad \boldsymbol{x} \ge 1, \text{ integer vector} \end{cases} \tag{8.9}$$

where $\overline{C}$ is the total capital.

8.5 Exercises

Problem 8.1. Design a hybrid intelligent algorithm to solve hybrid models for system reliability design problem (for example, some elements have fuzzy lifetimes and some have random lifetimes).

Problem 8.2. Build uncertain models for system reliability design problem (for example, the lifetimes of elements are uncertain variables with identification function (λ, ρ)), and design a hybrid intelligent algorithm to solve them.

Chapter 9
Project Scheduling Problem

Project scheduling problem is to determine the schedule of allocating resources so as to balance the total cost and the completion time. Uncertainty always exists in project scheduling problem due to the vagueness of project activity duration times. This chapter will introduce some optimization models for uncertain project scheduling problems.

9.1 Problem Description

Project scheduling is usually represented by a directed acyclic graph where nodes correspond to milestones, and arcs to activities which are basically characterized by the times and costs consumed.

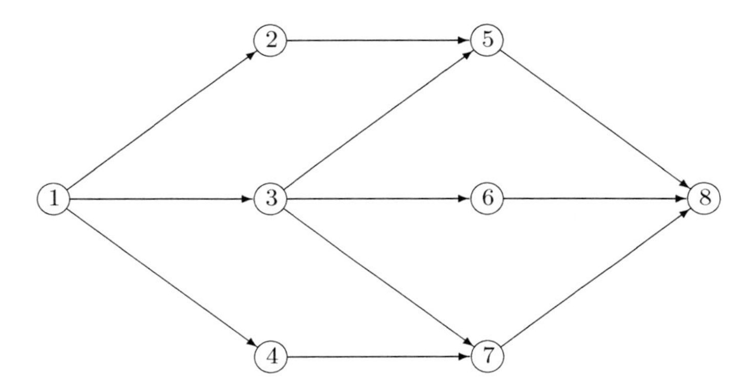

Fig. 9.1 A Project Graph

Let $\mathcal{G} = (\mathcal{V}, \mathcal{A})$ be a directed acyclic graph, where $\mathcal{V} = \{1, 2, \cdots, n, n+1\}$ is the set of nodes, $\mathcal{A}$ is the set of arcs, $(i, j) \in \mathcal{A}$ is the arc of the graph $\mathcal{G}$ from nodes i to j. It is well-known that we can rearrange the indexes of the nodes in $\mathcal{V}$ such that $i < j$ for all $(i, j) \in \mathcal{A}$.

B. Liu: Theory and Practice of Uncertain Programming, STUDFUZZ 239, pp. 139–146.
springerlink.com

Before we begin to study project scheduling problem with stochastic activity duration times, we first make some assumptions: (a) all of the costs needed are obtained via loans with some given interest rate; and (b) each activity can be processed only if the loan needed is allocated and all the foregoing activities are finished.

In order to model the project scheduling problem, we introduce the following indices and parameters:

ξ_{ij}: uncertain duration time of activity (i, j) in $\mathcal{A}$;

c_{ij}: cost of activity (i, j) in $\mathcal{A}$;

r: the interest rate;

x_i: integer decision variable representing the allocating time of all loans needed for all activities (i, j) in $\mathcal{A}$.

For simplicity, we also write $\boldsymbol{\xi} = \{\xi_{ij} : (i, j) \in \mathcal{A}\}$, $\boldsymbol{x} = (x_1, x_2, \cdots, x_n)$. Denote $T_i(\boldsymbol{x}, \boldsymbol{\xi})$ as the starting time of all activities (i, j) in $\mathcal{A}$. According to the assumptions, the starting time of the total project should be

$$T_1(\boldsymbol{x}, \boldsymbol{\xi}) = x_1. \tag{9.1}$$

The starting time of activities (i, j) in $\mathcal{A}$ should be

$$T_i(\boldsymbol{x}, \boldsymbol{\xi}) = x_i \vee \max_{(k,i)\in\mathcal{A}} \{T_k(\boldsymbol{x}, \boldsymbol{\xi}) + \xi_{ki}\}, \quad i = 2, 3, \cdots, n. \tag{9.2}$$

The completion time of the total project is

$$T(\boldsymbol{x}, \boldsymbol{\xi}) = \max_{(k,n+1)\in\mathcal{A}} \{T_k(\boldsymbol{x}, \boldsymbol{\xi}) + \xi_{k,n+1}\}. \tag{9.3}$$

Therefore, the total cost of the project can be written as

$$C(\boldsymbol{x}, \boldsymbol{\xi}) = \sum_{(i,j)\in\mathcal{A}} c_{ij} (1 + r)^{\lceil (T(\boldsymbol{x}, \boldsymbol{\xi}) - x_i \rceil} \tag{9.4}$$

where $\lceil a \rceil$ represents the minimal integer greater than or equal to a.

9.2 Stochastic Models

In this section, we assume that all duration times are all random variables, and introduce stochastic expected cost minimization model, α-cost minimization model and probability maximization model.

Stochastic Expected Cost Minimization Model

If we want to minimize the expected cost of the project under the expected completion time constraint, we may construct the following stochastic expected cost minimization model (Ke and Liu [123]),

$$\begin{cases} \min E[C(\boldsymbol{x}, \boldsymbol{\xi})] \\ \text{subject to:} \\ \qquad E[T(\boldsymbol{x}, \boldsymbol{\xi})] \le T^0 \\ \qquad \boldsymbol{x} \ge 0, \text{ integer vector} \end{cases} \tag{9.5}$$

where T^0 is the due date of the project, $T(\boldsymbol{x}, \boldsymbol{\xi})$ and $C(\boldsymbol{x}, \boldsymbol{\xi})$ are the completion time and total cost defined by (9.3) and (9.4), respectively.

Example 9.1. Now let us consider a project scheduling problem shown in Figure 9.1. The duration times and the costs needed for the relevant activities in the project are presented in Table 9.1, and the monthly interest rate $r = 0.006$ according to some practical case. Note that the activity duration times are assumed to be normally distributed random variables $\mathcal{N}(\mu, \sigma^2)$.

Table 9.1 Random Duration Times and Costs of Activities

Arc	Duration Time (Month)	Cost	Arc	Duration Time (Month)	Cost
(1,2)	$\mathcal{N}(9,3)$	1500	(1,3)	$\mathcal{N}(5,2)$	1800
(1,4)	$\mathcal{N}(10,3)$	430	(2,5)	$\mathcal{N}(6,2)$	1600
(3,5)	$\mathcal{N}(8,2)$	800	(3,6)	$\mathcal{N}(8,2)$	500
(3,7)	$\mathcal{N}(9,3)$	2000	(4,7)	$\mathcal{N}(6,1)$	2100
(5,8)	$\mathcal{N}(10,2)$	550	(6,8)	$\mathcal{N}(15,3)$	530
(7,8)	$\mathcal{N}(11,2)$	630			

Now we are requested to finish the project within 32 months, i.e., $T^0 = 32$ (month). A run of the hybrid intelligent algorithm (5000 cycles in simulation, 4000 generations in GA) shows that the expected cost $E[C(\boldsymbol{x}^*, \boldsymbol{\xi})] = 14345$, the expected completion time $E[T(\boldsymbol{x}^*, \boldsymbol{\xi})] = 30.4$, and the optimal schedule for the project is shown in Table 9.2.

Table 9.2 Schedule of Expected Cost Model

Date	1	6	12	13	17	18
Node	1	3	2,4	6	7	5
Loan	3730	3300	3700	530	630	550

Stochastic α-Cost Minimization Model

The α-cost of a project is defined as $\min\left\{\bar{C} \mid \Pr\{C(\boldsymbol{x}, \boldsymbol{\xi}) \le \bar{C}\} \ge \alpha\right\}$, where α is a predetermined confidence level. If we want to minimize the α-cost of the project under the completion time chance constraint with a predetermined confidence level β, we have the following stochastic α-cost minimization model (Ke and Liu [123]),

$$\begin{cases} \min \bar{C} \\ \text{subject to:} \\ \qquad \Pr\{C(\boldsymbol{x},\boldsymbol{\xi}) \le \bar{C}\} \ge \alpha \\ \qquad \Pr\{T(\boldsymbol{x},\boldsymbol{\xi}) \le T^0\} \ge \beta \\ \qquad \boldsymbol{x} \ge 0, \text{ integer vector} \end{cases} \tag{9.6}$$

where T^0 is the due date of the project, $T(\boldsymbol{x},\boldsymbol{\xi})$ and $C(\boldsymbol{x},\boldsymbol{\xi})$ are the completion time and total cost defined by (9.3) and (9.4), respectively.

Example 9.2. If $\alpha = 0.9$, $\beta = 0.9$ and $T^0 = 32$ (month), then a run of the hybrid intelligent algorithm (5000 cycles in simulation, 4000 generations in GA) shows that the α-cost $C^* = 14502$, $\Pr\{C(\boldsymbol{x}^*,\boldsymbol{\xi}) \le C^*\} = 0.901$, $\Pr\{T(\boldsymbol{x}^*,\boldsymbol{\xi}) \le T^0\} = 0.905$, and the optimal schedule is presented in Table 9.3.

Table 9.3 Schedule of α-Cost Minimization Model

Date	1	5	12	13	18
Node	1	3	2,4	6	5,7
Loan	3730	3300	3700	530	1180

Probability Maximization Model

If we want to maximize the probability that the total cost should not exceed a predetermined level C^0 subject to the chance constraint $\Pr\{T(\boldsymbol{x},\boldsymbol{\xi}) \le T^0\} \ge \beta$, then we have the following probability maximization model (Ke and Liu [123]),

$$\begin{cases} \max \Pr\left\{C(\boldsymbol{x},\boldsymbol{\xi}) \le C^0\right\} \\ \text{subject to:} \\ \qquad \Pr\{T(\boldsymbol{x},\boldsymbol{\xi}) \le T^0\} \ge \beta \\ \qquad \boldsymbol{x} \ge 0, \text{ integer vector} \end{cases} \tag{9.7}$$

where $T(\boldsymbol{x},\boldsymbol{\xi})$ and $C(\boldsymbol{x},\boldsymbol{\xi})$ are the completion time and total cost defined by (9.3) and (9.4), respectively.

Example 9.3. If $C^0 = 14530$, $T^0 = 32$ and $\beta = 0.9$, then a run of the hybrid intelligent algorithm (5000 cycles in simulation, 4000 generations in GA) shows that the probability $\Pr\{C(\boldsymbol{x}^*,\boldsymbol{\xi}) \le C^0\} = 0.912$, $\Pr\{T(\boldsymbol{x}^*,\boldsymbol{\xi}) \le T^0\} = 0.906$ and the optimal schedule is presented in Table 9.4.

Table 9.4 Schedule of Probability Maximization Model

Date	1	6	10	11	13	17	18
Node	1	3	2	4	6	7	5
Loan	3730	3300	1600	2100	530	630	550

9.3 Fuzzy Models

This section will assume that the activity duration times are all fuzzy variables rather than random ones, and introduce fuzzy expected cost minimization model, α-cost minimization model and credibility maximization model.

Fuzzy Expected Cost Minimization Model

If we want to minimize the expected cost of the project under the expected completion time constraint, we may construct the following fuzzy expected cost minimization model (Ke and Liu [125]),

$$\begin{cases} \min E[C(\boldsymbol{x}, \boldsymbol{\xi})] \\ \text{subject to:} \\ \qquad E[T(\boldsymbol{x}, \boldsymbol{\xi})] \leq T^0 \\ \qquad \boldsymbol{x} \geq 0, \text{ integer vector} \end{cases} \tag{9.8}$$

where T^0 is the due date of the project, $T(\boldsymbol{x}, \boldsymbol{\xi})$ and $C(\boldsymbol{x}, \boldsymbol{\xi})$ are the completion time and total cost defined by (9.3) and (9.4), respectively.

Example 9.4. Now let us consider a project scheduling problem shown in Figure 9.1. The duration times and the costs needed for the relevant activities in the project are presented in Table 9.5, and the monthly interest rate $r = 0.006$. Note that the activity duration times are assumed to be triangular fuzzy variables.

Table 9.5 Fuzzy Duration Times and Costs of Activities

Arc	Duration Time (Month)	Cost	Arc	Duration Time (Month)	Cost
(1,2)	(7, 9, 12)	1500	(1,3)	(3,5,7)	1800
(1,4)	(7,10,12)	430	(2,5)	(4,6,9)	1600
(3,5)	(6,8,11)	800	(3,6)	(6,8,10)	500
(3,7)	(6,9,12)	2000	(4,7)	(5,6,8)	2100
(5,8)	(8,10,12)	550	(6,8)	(13,16,18)	530
(7,8)	(9,11,13)	630			

Table 9.6 Schedule of Expected Cost Model

Date	1	7	12	13	14	18	19
Node	1	3	6	4	2	7	5
Loan	3730	3300	530	2100	1600	630	550

Now we are requested to finish the project within 32 months, i.e., $T^0 = 32$ (month). A run of the hybrid intelligent algorithm (5000 cycles in simulation, 4000 generations in GA) shows that the expected cost $E[C(\boldsymbol{x}^*, \boldsymbol{\xi})] = 14286$, the expected completion time $E[T(\boldsymbol{x}^*, \boldsymbol{\xi})] = 30.4$, and the optimal schedule for the project is shown in Table 9.6.

Fuzzy α-Cost Minimization Model

The α-cost of a project is defined as $\min\left\{\bar{C}|\text{Cr}\{C(\boldsymbol{x}, \boldsymbol{\xi}) \le \bar{C}\} \ge \alpha\right\}$, where α is a predetermined confidence level. If we want to minimize the α-cost of the project under the completion time chance constraint with a predetermined confidence level β, we have the following fuzzy α-cost minimization model (Ke and Liu [125]),

$$\begin{cases} \min \bar{C} \\ \text{subject to:} \\ \qquad \text{Cr}\{C(\boldsymbol{x}, \boldsymbol{\xi}) \le \bar{C}\} \ge \alpha \\ \qquad \text{Cr}\{T(\boldsymbol{x}, \boldsymbol{\xi}) \le T^0\} \ge \beta \\ \qquad \boldsymbol{x} \ge 0, \text{ integer vector} \end{cases} \tag{9.9}$$

where T^0 is the due date of the project, $T(\boldsymbol{x}, \boldsymbol{\xi})$ and $C(\boldsymbol{x}, \boldsymbol{\xi})$ are the completion time and total cost defined by (9.3) and (9.4), respectively.

Example 9.5. If $\alpha = 0.9$, $\beta = 0.9$ and $T^0 = 32$ (month), then a run of the hybrid intelligent algorithm (5000 cycles in simulation, 4000 generations in GA) shows that the α-cost $C^* = 14331$, $\text{Cr}\{C(\boldsymbol{x}^*, \boldsymbol{\xi}) \le C^*\} = 0.913$, $\text{Cr}\{T(\boldsymbol{x}^*, \boldsymbol{\xi}) \le T^0\} = 0.917$, and the optimal schedule is presented in Table 9.7.

Table 9.7 Schedule of α-Cost Minimization Model

Date	1	6	13	14	15	20	21
Node	1	3	4	2	6	7	5
Loan	3730	3300	2100	1600	530	630	550

Credibility Maximization Model

If we want to maximize the credibility that the total cost should not exceed a predetermined level C^0 subject to the chance constraint $\text{Cr}\{T(\boldsymbol{x}, \boldsymbol{\xi})$

$\leq T^0\} \geq \beta$, then we have the following credibility maximization model (Ke and Liu [125]),

$$\begin{cases} \max \mathrm{Cr}\left\{C(\boldsymbol{x}, \boldsymbol{\xi}) \leq C^0\right\} \\ \text{subject to:} \\ \qquad \mathrm{Cr}\{T(\boldsymbol{x}, \boldsymbol{\xi}) \leq T^0\} \geq \beta \\ \qquad \boldsymbol{x} \geq 0, \text{ integer vector} \end{cases} \tag{9.10}$$

where $T(\boldsymbol{x}, \boldsymbol{\xi})$ and $C(\boldsymbol{x}, \boldsymbol{\xi})$ are the completion time and total cost defined by (9.3) and (9.4), respectively.

Example 9.6. If $C^0 = 14370$, $T^0 = 32$ and $\beta = 0.9$, then a run of the hybrid intelligent algorithm (5000 cycles in simulation, 4000 generations in GA) shows that the credibility $\mathrm{Cr}\{C(\boldsymbol{x}^*, \boldsymbol{\xi}) \leq C^0\} = 0.95$, $\mathrm{Cr}\{T(\boldsymbol{x}^*, \boldsymbol{\xi}) \leq T^0\} = 0.95$ and the optimal schedule is presented in Table 9.8.

Table 9.8 Schedule of Credibility Maximization Model

Date	1	6	13	14	19
Node	1	3	4,6	2	5,7
Loan	3730	3300	2630	1600	1180

9.4 Hybrid Models

We suppose that the duration times are hybrid variables, and introduce hybrid expected cost minimization model, α-cost minimization model and chance maximization model.

Hybrid Expected Cost Minimization Model

If we want to minimize the expected cost of the project under the expected completion time constraint, we may construct the following hybrid expected cost minimization model,

$$\begin{cases} \min E[C(\boldsymbol{x}, \boldsymbol{\xi})] \\ \text{subject to:} \\ \qquad E[T(\boldsymbol{x}, \boldsymbol{\xi})] \leq T^0 \\ \qquad \boldsymbol{x} \geq 0, \text{ integer vector} \end{cases} \tag{9.11}$$

where T^0 is the due date of the project, $T(\boldsymbol{x}, \boldsymbol{\xi})$ and $C(\boldsymbol{x}, \boldsymbol{\xi})$ are the completion time and total cost defined by (9.3) and (9.4), respectively.

Hybrid α-Cost Minimization Model

The α-cost of a project is defined as $\min\left\{\bar{C} | \mathrm{Ch}\{C(\boldsymbol{x}, \boldsymbol{\xi}) \leq \bar{C}\} \geq \alpha\right\}$, where α is a predetermined confidence level. If we want to minimize the α-cost of the project under the completion time chance constraint with predetermined confidence level β, we have the following hybrid α-cost minimization model (Ke and Liu [124]),

$$\begin{cases} \min \bar{C} \\ \text{subject to:} \\ \qquad \text{Ch}\{C(\boldsymbol{x}, \boldsymbol{\xi}) \le \bar{C}\} \ge \alpha \\ \qquad \text{Ch}\{T(\boldsymbol{x}, \boldsymbol{\xi}) \le T^0\} \ge \beta \\ \qquad \boldsymbol{x} \ge 0, \text{ integer vector} \end{cases} \tag{9.12}$$

where T^0 is the due date of the project, $T(\boldsymbol{x}, \boldsymbol{\xi})$ and $C(\boldsymbol{x}, \boldsymbol{\xi})$ are the completion time and total cost defined by (9.3) and (9.4), respectively.

Chance Maximization Model

If we want to maximize the chance that the total cost should not exceed a predetermined level C^0 subject to the chance constraint $\text{Ch}\{T(\boldsymbol{x}, \boldsymbol{\xi}) \le T^0\} \ge \alpha$, then we have the following chance maximization model,

$$\begin{cases} \max \text{Ch}\left\{C(\boldsymbol{x}, \boldsymbol{\xi}) \le C^0\right\} \\ \text{subject to:} \\ \qquad \text{Ch}\{T(\boldsymbol{x}, \boldsymbol{\xi}) \le T^0\} \ge \alpha \\ \qquad \boldsymbol{x} \ge 0, \text{ integer vector} \end{cases} \tag{9.13}$$

where $T(\boldsymbol{x}, \boldsymbol{\xi})$ and $C(\boldsymbol{x}, \boldsymbol{\xi})$ are the completion time and total cost defined by (9.3) and (9.4), respectively.

9.5 Exercises

Problem 9.1. Design a hybrid intelligent algorithm to solve hybrid models for project scheduling problem (for example, the duration times are random and costs are fuzzy).

Problem 9.2. Build uncertain models for project scheduling problem (for example, the duration times are uncertain variables with identification function (λ, ρ)), and design a hybrid intelligent algorithm to solve them.

Chapter 10
Vehicle Routing Problem

Vehicle routing problem (VRP) is concerned with finding efficient routes, beginning and ending at a central depot, for a fleet of vehicles to serve a number of customers. See Figure 10.1.

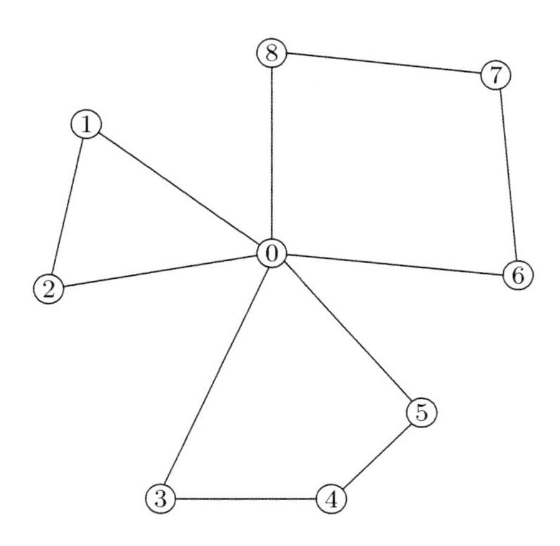

Fig. 10.1 A Vehicle Routing Graph

Due to its wide applicability and economic importance, VRP has been extensively studied. Practically, there are uncertain factors in VRP, such as demands of customers, travel times between customers, customers to be visited, locations of customers, capacities of vehicles, and number of vehicles available. This fact provides a motivation to study uncertain VRP. This chapter introduces some typical models for VRP.

10.1 Problem Description

We assume that: (a) a vehicle will be assigned for only one route on which there may be more than one customer; (b) a customer will be visited by one and only one vehicle; (c) each route begins and ends at the depot; and (d) each

B. Liu: Theory and Practice of Uncertain Programming, STUDFUZZ 239, pp. 147–155.
springerlink.com

customer specifies its time window within which the delivery is permitted or preferred to start.

Let us first introduce the following indices and model parameters:

$i = 0$: depot;

$i = 1, 2, \cdots, n$: customers;

$k = 1, 2, \cdots, m$: vehicles;

D_{ij}: the travel distance from customers i to j, $i, j = 0, 1, 2, \cdots, n$;

T_{ij}: the uncertain travel time from customers i to j, $i, j = 0, 1, 2, \cdots, n$;

S_i: the unloading time at customer i, $i = 1, 2, \cdots, n$;

$[a_i, b_i]$: the time window of customer i, where a_i and b_i are the beginning and end of the time window, $i = 1, 2, \cdots, n$, respectively.

In this book, the operational plan is represented by Liu's formulation [181] via three decision vectors $\boldsymbol{x}$, $\boldsymbol{y}$ and $\boldsymbol{t}$, where

$\boldsymbol{x} = (x_1, x_2, \cdots, x_n)$: integer decision vector representing n customers with $1 \le x_i \le n$ and $x_i \ne x_j$ for all $i \ne j$, $i, j = 1, 2, \cdots, n$. That is, the sequence $\{x_1, x_2, \cdots, x_n\}$ is a rearrangement of $\{1, 2, \cdots, n\}$;

$\boldsymbol{y} = (y_1, y_2, \cdots, y_{m-1})$: integer decision vector with $y_0 \equiv 0 \le y_1 \le y_2 \le \cdots \le y_{m-1} \le n \equiv y_m$;

$\boldsymbol{t} = (t_1, t_2, \cdots, t_m)$: each t_k represents the starting time of vehicle k at the depot, $k = 1, 2, \cdots, m$.

We note that the operational plan is fully determined by the decision vectors $\boldsymbol{x}$, $\boldsymbol{y}$ and $\boldsymbol{t}$ in the following way. For each k ($1 \le k \le m$), if $y_k = y_{k-1}$, then vehicle k is not used; if $y_k > y_{k-1}$, then vehicle k is used and starts from the depot at time t_k, and the tour of vehicle k is $0 \to x_{y_{k-1}+1} \to x_{y_{k-1}+2} \to \cdots \to x_{y_k} \to 0$. Thus the tours of all vehicles are as follows:

$$\begin{aligned}
&\text{Vehicle 1: } 0 \to x_{y_0+1} \to x_{y_0+2} \to \cdots \to x_{y_1} \to 0; \\
&\text{Vehicle 2: } 0 \to x_{y_1+1} \to x_{y_1+2} \to \cdots \to x_{y_2} \to 0; \\
&\quad \cdots \\
&\text{Vehicle } m\text{: } 0 \to x_{y_{m-1}+1} \to x_{y_{m-1}+2} \to \cdots \to x_{y_m} \to 0.
\end{aligned} \tag{10.1}$$

It is clear that this type of representation is intuitive, and the total number of decision variables is $n + 2m - 1$. We also note that the above decision variables $\boldsymbol{x}$, $\boldsymbol{y}$ and $\boldsymbol{t}$ ensure that: (a) each vehicle will be used at most one time; (b) all tours begin and end at the depot; (c) each customer will be visited by one and only one vehicle; and (d) there is no subtour.

Let $f_i(\boldsymbol{x}, \boldsymbol{y}, \boldsymbol{t})$ be the arrival time function of some vehicles at customers i for $i = 1, 2, \cdots, n$. We remind readers that $f_i(\boldsymbol{x}, \boldsymbol{y}, \boldsymbol{t})$ are determined by the decision variables $\boldsymbol{x}$, $\boldsymbol{y}$ and $\boldsymbol{t}$, $i = 1, 2, \cdots, n$. Since unloading can start either immediately, or later, when a vehicle arrives at a customer, the calculation of $f_i(\boldsymbol{x}, \boldsymbol{y}, \boldsymbol{t})$ is heavily dependent on the operational strategy. Here we assume that the customer does not permit a delivery earlier than the time window. That is, the vehicle will wait to unload until the beginning of the time window if it arrives before the time window. If a vehicle arrives at a customer after the beginning of the time window, unloading will start immediately. For each k with $1 \le k \le m$, if vehicle k is used (i.e., $y_k > y_{k-1}$), then we have

$$f_{x_{y_{k-1}+1}}(\boldsymbol{x},\boldsymbol{y},\boldsymbol{t}) = t_k + T_{0x_{y_{k-1}+1}} \tag{10.2}$$

and

$$\begin{aligned} f_{x_{y_{k-1}+j}}(\boldsymbol{x},\boldsymbol{y},\boldsymbol{t}) =& f_{x_{y_{k-1}+j-1}}(\boldsymbol{x},\boldsymbol{y},\boldsymbol{t}) \vee a_{x_{y_{k-1}+j-1}} \\ &+ S_{x_{y_{k-1}+j-1}} + T_{x_{y_{k-1}+j-1}x_{y_{k-1}+j}} \end{aligned} \tag{10.3}$$

for $2 \le j \le y_k - y_{k-1}$. It follows from the uncertainty of travel times T_{ij}'s that the arrival times $f_i(\boldsymbol{x},\boldsymbol{y},\boldsymbol{t})$, $i = 1, 2, \cdots, n$ are uncertain variables fully determined by (10.2) and (10.3).

Let $g(\boldsymbol{x},\boldsymbol{y})$ be the total travel distance of all vehicles. Then we have

$$g(\boldsymbol{x},\boldsymbol{y}) = \sum_{k=1}^{m} g_k(\boldsymbol{x},\boldsymbol{y}) \tag{10.4}$$

where

$$g_k(\boldsymbol{x},\boldsymbol{y}) = \begin{cases} D_{0x_{y_{k-1}+1}} + \sum\limits_{j=y_{k-1}+1}^{y_k-1} D_{x_j x_{j+1}} + D_{x_{y_k}0}, & \text{if } y_k > y_{k-1} \\ 0, & \text{if } y_k = y_{k-1} \end{cases}$$

for $k = 1, 2, \cdots, m$.

10.2 Stochastic Models

Now we assume that the travel times are random variables, and introduce stochastic distance minimization model and probability maximization model.

Stochastic Distance Minimization Model

If we hope that all customers are visited within their time windows with a confidence level α, then we have the following chance constraint,

$$\Pr\{a_i \le f_i(\boldsymbol{x},\boldsymbol{y},\boldsymbol{t}) \le b_i, i = 1, 2, \cdots, n\} \ge \alpha. \tag{10.5}$$

If we want to minimize the total travel distance of all vehicles subject to the time window constraint, then we have the following stochastic distance minimization model (Liu and Lai [183]),

$$\begin{cases} \min g(\boldsymbol{x},\boldsymbol{y}) \\ \text{subject to:} \\ \qquad \Pr\{a_i \le f_i(\boldsymbol{x},\boldsymbol{y},\boldsymbol{t}) \le b_i, i = 1, 2, \cdots, n\} \ge \alpha \\ \qquad 1 \le x_i \le n, \quad i = 1, 2, \cdots, n \\ \qquad x_i \ne x_j, \quad i \ne j, \ i, j = 1, 2, \cdots, n \\ \qquad 0 \le y_1 \le y_2 \le \cdots \le y_{m-1} \le n \\ \qquad x_i, y_j, \quad i = 1, 2, \cdots, n, \quad j = 1, 2, \cdots, m-1, \quad \text{integers.} \end{cases} \tag{10.6}$$

Hybrid Intelligent Algorithm

In order to solve the stochastic models, we may employ the hybrid intelligent algorithm documented in Chapter 4 provided that the representation structure, initialization, crossover and mutation operations are revised as follows.

We represent an operational plan by the chromosome $V = (\boldsymbol{x}, \boldsymbol{y}, \boldsymbol{t})$, where the genes $\boldsymbol{x}, \boldsymbol{y}, \boldsymbol{t}$ are the same as the decision vectors. Without loss of generality, we also assume that the time window at the depot is $[a, b]$. This means that the gene $\boldsymbol{t}$ will be restricted in the hypercube $[a, b]^m$.

Let us show how to initialize a chromosome randomly. For gene $\boldsymbol{x}$, we define a sequence $\{x_1, x_2, \cdots, x_n\}$ with $x_i = i$, $i = 1, 2, \cdots, n$, and repeat the following process from $j = 1$ to n: generating a random position n' between j and n, and exchanging the values of x_j and $x_{n'}$. It is clear that $\{x_1, x_2, \cdots, x_n\}$ is just a random rearrangement of $\{1, 2, \cdots, n\}$. Then we obtain a gene $\boldsymbol{x} = (x_1, x_2, \cdots, x_n)$. For each i with $1 \le i \le m-1$, we set y_i as a random integer between 0 and n. Then we rearrange the sequence $\{y_1, y_2, \cdots, y_{m-1}\}$ from small to large. We thus have a gene $\boldsymbol{y} = (y_1, y_2, \cdots, y_{m-1})$. Finally, for each i with $1 \le i \le m$, we set t_i as a random number on the time window $[a, b]$. Then we get a gene $\boldsymbol{t} = (t_1, t_2, \cdots, t_m)$. If the generated chromosome $V = (\boldsymbol{x}, \boldsymbol{y}, \boldsymbol{t})$ is proven to be feasible, then it is accepted as a chromosome; otherwise we repeat the above process until a feasible chromosome is obtained.

Let us illustrate the crossover operator on the pair V_1 and V_2. We denote $V_1 = (\boldsymbol{x}_1, \boldsymbol{y}_1, \boldsymbol{t}_1)$ and $V_2 = (\boldsymbol{x}_2, \boldsymbol{y}_2, \boldsymbol{t}_2)$. First, we generate a random number c from the open interval $(0, 1)$ and define

$$\boldsymbol{t}_1' = c \cdot \boldsymbol{t}_1 + (1-c) \cdot \boldsymbol{t}_2, \quad \boldsymbol{t}_2' = (1-c) \cdot \boldsymbol{t}_1 + c \cdot \boldsymbol{t}_2.$$

The two children V_1' and V_2' are produced by the crossover operation as follows: $V_1' = (\boldsymbol{x}_1, \boldsymbol{y}_2, \boldsymbol{t}_1')$ and $V_2' = (\boldsymbol{x}_2, \boldsymbol{y}_1, \boldsymbol{t}_2')$.

We mutate the chromosome $V = (\boldsymbol{x}, \boldsymbol{y}, \boldsymbol{t})$ in the following way. For the gene $\boldsymbol{x}$, we randomly generate two mutation positions n_1 and n_2 between 1 and n, and rearrange the sequence $\{x_{n_1}, x_{n_1+1}, \cdots, x_{n_2}\}$ at random to form a new sequence $\{x_{n_1}', x_{n_1+1}', \cdots, x_{n_2}'\}$. We thus obtain a new gene

$$\boldsymbol{x}' = (x_1, \cdots, x_{n_1-1}, x_{n_1}', x_{n_1+1}', \cdots, x_{n_2}', x_{n_2+1}, \cdots, x_n).$$

Similarly, for gene $\boldsymbol{y}$, we generate two random mutation positions n_1 and n_2 between 1 and $m-1$, and set y_i as a random integer number y_i' between 0 and n for $i = n_1, n_1+1, \cdots, n_2$. We then rearrange the sequence

$$\{y_1, \cdots, y_{n_1-1}, y_{n_1}', y_{n_1+1}', \cdots, y_{n_2}', y_{n_2+1}, \cdots, y_{m-1}\}$$

from small to large and obtain a new gene $\boldsymbol{y}'$. For the gene $\boldsymbol{t}$, we choose a mutation direction $\boldsymbol{d}$ in $\Re^m$ randomly. If $\boldsymbol{t} + M \cdot \boldsymbol{d}$ is not in the time window $[a, b]^m$, then we set M as a random number between 0 and M until it is in $[a, b]^m$, where M is a predetermined step length. If the above process cannot

Table 10.1 Travel Distance Matrix

LCTs	0	1	2	3	4	5	6	7
1	18							
2	14	20						
3	14	34	15					
4	21	55	41	28				
5	17	49	43	36	21			
6	21	57	55	51	36	16		
7	18	49	52	51	43	22	13	
8	14	22	35	44	55	41	43	32

Table 10.2 Random Travel Time Matrix (μ, σ^2)

LCTs	0	1	2	3
1	(50,25)			
2	(10,5)	(40,20)		
3	(50,25)	(10,5)	(40,20)	
4	(50,25)	(35,17)	(35,17)	(30,15)
5	(50,25)	(15,7)	(40,20)	(5,2)
6	(15,7)	(40,20)	(10,5)	(45,22)
7	(50,25)	(15,7)	(45,22)	(10,5)
8	(50,25)	(10,5)	(35,17)	(30,15)
LCTs	**4**	**5**	**6**	**7**
5	(30,15)			
6	(35,17)	(40,20)		
7	(30,15)	(10,5)	(40,20)	
8	(10,5)	(30,15)	(35,17)	(35,17)

yield a gene $\boldsymbol{t}$ in $[a, b]^m$ in a predetermined number of iterations, then we set $M = 0$. We replace the parent gene $\boldsymbol{t}$ with its child $\boldsymbol{t}' = \boldsymbol{t} + M \cdot \boldsymbol{d}$.

Example 10.1. We assume that there are 8 customers labeled "$1, 2, \cdots, 8$" in a company and one depot labeled "0". We assume that the travel distances among the depot and customers are listed in Table 10.1.

The travel times among the depot and customers are all normally distributed variables $\mathcal{N}(\mu, \sigma^2)$, which are given in Table 10.2.

The time windows of customers are shown in Table 10.3.

We suppose that the unloading times ($S_i, i = 1, 2, \cdots, 8$) at locations are all 15 minutes.

We assign a confidence level $\alpha = 0.80$ at which all customers are visited within their time windows. If we want to minimize the total travel distance of all vehicles subject to the chance constraint, then we have a stochastic distance minimization model. A run of the hybrid intelligent algorithm (10000

Table 10.3 Time Windows of Customers

i	$[a_i, b_i]$	i	$[a_i, b_i]$	i	$[a_i, b_i]$
1	$[09:30, 14:10]$	2	$[09:20, 11:00]$	3	$[09:40, 11:10]$
4	$[09:20, 13:00]$	5	$[09:10, 15:20]$	6	$[08:20, 10:00]$
7	$[09:40, 12:10]$	8	$[09:20, 10:00]$		

cycles in simulation, 5000 generations in GA) shows that the best operational plan is:

Vehicle 1: depot$\rightarrow$ 6 $\rightarrow$ 7 $\rightarrow$depot, starting time = 8:45;
Vehicle 2: depot$\rightarrow$ 3 $\rightarrow$depot, starting time = 9:17;
Vehicle 3: depot$\rightarrow$ 8 $\rightarrow$ 1 $\rightarrow$ 2 $\rightarrow$ 5 $\rightarrow$ 4 $\rightarrow$depot, starting time = 8:35.

The total travel distance of the three vehicles is 221. Furthermore, when the obtained operational plan is performed, we have

$$\Pr\{a_i \le f_i(\boldsymbol{x}^*, \boldsymbol{y}^*, \boldsymbol{t}^*) \le b_i, i = 1, 2, \cdots, 8\} = 0.85.$$

Probability Maximization Model

If we hope that total travel distance does not exceed a fixed number $\overline{g}$, then we have a distance constraint $g(\boldsymbol{x}, \boldsymbol{y}) \le \overline{g}$. If we want to maximize the probability that all customers are visited within their time windows subject to the distance constraint, then we have the following probability maximization model,

$$\begin{cases} \max \Pr\{a_i \le f_i(\boldsymbol{x}, \boldsymbol{y}, \boldsymbol{t}) \le b_i, i = 1, 2, \cdots, n\} \\ \text{subject to:} \\ \qquad g(\boldsymbol{x}, \boldsymbol{y}) \le \overline{g} \\ \qquad 1 \le x_i \le n, \quad i = 1, 2, \cdots, n \\ \qquad x_i \ne x_j, \quad i \ne j, \ i, j = 1, 2, \cdots, n \\ \qquad 0 \le y_1 \le y_2 \le \cdots \le y_{m-1} \le n \\ \qquad x_i, y_j, \quad i = 1, 2, \cdots, n, \quad j = 1, 2, \cdots, m-1, \quad \text{integers.} \end{cases} \tag{10.7}$$

Example 10.2. We set $\overline{g} = 240$. A run of the hybrid intelligent algorithm (10000 cycles in simulation, 5000 generations in GA) shows that the best operational plan is:

Vehicle 1: depot$\rightarrow$ 1 $\rightarrow$ 3 $\rightarrow$ 2 $\rightarrow$depot, starting time = 8:51;
Vehicle 2: depot$\rightarrow$ 6 $\rightarrow$ 5 $\rightarrow$ 7 $\rightarrow$ 4 $\rightarrow$depot, starting time = 9:04;
Vehicle 3: depot$\rightarrow$ 8 $\rightarrow$depot, starting time = 8:58.

When the obtained operational plan is performed, the total travel distance is 232, and

$$\Pr\{a_i \le f_i(\boldsymbol{x}^*, \boldsymbol{y}^*, \boldsymbol{t}^*) \le b_i, i = 1, 2, \cdots, 8\} = 0.88.$$

10.3 Fuzzy Models

Here we assume that the travel times are fuzzy variables instead of stochastic variables. Since the travel times are fuzzy variables, every customer will be visited at a fuzzy time.

Fuzzy Distance Minimization Model

If we hope that all customers are visited within their time windows with a confidence level α, then we have the following chance constraint,

$$\text{Cr}\{a_i \leq f_i(\boldsymbol{x}, \boldsymbol{y}, \boldsymbol{t}) \leq b_i, i = 1, 2, \cdots, n\} \geq \alpha. \tag{10.8}$$

If we want to minimize the total distance traveled of all vehicles subject to time window constraints, then we have the following fuzzy distance minimization model (Zheng and Liu [336]),

$$\begin{cases} \min g(\boldsymbol{x}, \boldsymbol{y}) \\ \text{subject to:} \\ \qquad \text{Cr}\{a_i \leq f_i(\boldsymbol{x}, \boldsymbol{y}, \boldsymbol{t}) \leq b_i, i = 1, 2, \cdots, n\} \geq \alpha \\ \qquad 1 \leq x_i \leq n, \quad i = 1, 2, \cdots, n \\ \qquad x_i \neq x_j, \quad i \neq j, \ i, j = 1, 2, \cdots, n \\ \qquad 0 \leq y_1 \leq y_2 \leq \cdots \leq y_{m-1} \leq n \\ \qquad x_i, y_j, \quad i = 1, 2, \cdots, n, \quad j = 1, 2, \cdots, m-1, \quad \text{integers.} \end{cases}$$

Example 10.3. Let us consider a fuzzy vehicle routing problem shown in Figure 10.1. We assume that the distance matrix is listed in Table 10.1 and the time windows customers are given in Table 10.3. We also assume that

Table 10.4 Fuzzy Travel Time Matrix

LCTs	0	1	2	3
1	(25,50,75)			
2	(5,10,15)	(20,40,60)		
3	(25,50,75)	(5,10,15)	(20,40,60)	
4	(25,50,75)	(17,35,53)	(17,35,53)	(15,30,45)
5	(25,50,75)	(7,15,23)	(20,40,60)	(2,5,8)
6	(7,15,23)	(20,40,60)	(5,10,15)	(22,45,68)
7	(25,50,75)	(7,15,23)	(22,45,68)	(5,10,15)
8	(25,50,75)	(5,10,15)	(17,35,53)	(15,30,45)
LCTs	**4**	**5**	**6**	**7**
5	(15,30,45)			
6	(17,35,53)	(20,40,60)		
7	(15,30,45)	(5,10,15)	(20,40,60)	
8	(5,10,15)	(15,30,45)	(17,35,53)	(17,35,53)

the travel times among the deport and customers are all triangular fuzzy variables as shown in Table 10.4. Finally, we suppose that the unloading times at the 8 locations are all 15 minutes.

If the confidence level α is 0.80, then a run of the hybrid intelligent algorithm (10000 cycles in simulation, 5000 generations in GA) shows that the best operational plan is:

Vehicle 1: depot$\to$ 6 $\to$ 7 $\to$ 4 $\to$ 5 $\to$depot, starting time = 9:44;
Vehicle 2: depot$\to$ 8 $\to$depot, starting time = 8:48;
Vehicle 3: depot$\to$ 2 $\to$ 3 $\to$ 1 $\to$depot, starting time = 9:21.

The total distance travelled by the three vehicles is 224. Furthermore, when the operational plan is performed, we have

$$\text{Cr}\left\{a_i \leq f_i(\boldsymbol{x}^*, \boldsymbol{y}^*, \boldsymbol{t}^*) \leq b_i, i = 1, 2, \cdots, 8\right\} = 0.87.$$

Credibility Maximization Model

If we hope that total travel distance does not exceed a fixed number $\overline{g}$, then we have a distance constraint $g(\boldsymbol{x}, \boldsymbol{y}) \leq \overline{g}$. If we want to maximize the credibility that all customers are visited within their time windows subject to the distance constraint, then we have the following credibility maximization model,

$$\begin{cases} \max \text{Cr}\left\{a_i \leq f_i(\boldsymbol{x}, \boldsymbol{y}, \boldsymbol{t}) \leq b_i, i = 1, 2, \cdots, n\right\} \\ \text{subject to:} \\ \qquad g(\boldsymbol{x}, \boldsymbol{y}) \leq \overline{g} \\ \qquad 1 \leq x_i \leq n, \quad i = 1, 2, \cdots, n \\ \qquad x_i \neq x_j, \quad i \neq j, \ i, j = 1, 2, \cdots, n \\ \qquad 0 \leq y_1 \leq y_2 \leq \cdots \leq y_{m-1} \leq n \\ \qquad x_i, y_j, \quad i = 1, 2, \cdots, n, \quad j = 1, 2, \cdots, m-1, \quad \text{integers.} \end{cases} \tag{10.9}$$

Example 10.4. We set $\overline{g} = 240$. A run of the hybrid intelligent algorithm (10000 cycles in simulation, 5000 generations in GA) shows that the best operational plan is

Vehicle 1: depot$\to$ 8 $\to$ 1 $\to$ 3 $\to$depot, starting time = 8:55;
Vehicle 2: depot$\to$ 6 $\to$ 5 $\to$ 4 $\to$ 7 $\to$depot, starting time = 8:51;
Vehicle 3: depot$\to$ 2 $\to$depot, starting time = 9:21.

When the optimal operational plan is performed, the total travel distance is 231, and

$$\text{Cr}\left\{a_i \leq f_i(\boldsymbol{x}^*, \boldsymbol{y}^*, \boldsymbol{t}^*) \leq b_i, i = 1, 2, \cdots, 8\right\} = 0.96.$$

10.4 Hybrid Models

Now we suppose that the travel times are hybrid variables, and introduce hybrid distance minimization model and chance maximization model.

Hybrid Distance Minimization Model

If we hope that all customers are visited within their time windows with confidence level α, then we have the following chance constraint,

$$\text{Ch}\{a_i \le f_i(\boldsymbol{x}, \boldsymbol{y}, \boldsymbol{t}) \le b_i, i = 1, 2, \cdots, n\} \ge \alpha. \tag{10.10}$$

If we want to minimize the total travel distance of all vehicles subject to the time window constraint, then we have the following hybrid distance minimization model,

$$\begin{cases} \min g(\boldsymbol{x}, \boldsymbol{y}) \\ \text{subject to:} \\ \qquad \text{Ch}\{a_i \le f_i(\boldsymbol{x}, \boldsymbol{y}, \boldsymbol{t}) \le b_i, i = 1, 2, \cdots, n\} \ge \alpha \\ \qquad 1 \le x_i \le n, \quad i = 1, 2, \cdots, n \\ \qquad x_i \ne x_j, \quad i \ne j, \ i, j = 1, 2, \cdots, n \\ \qquad 0 \le y_1 \le y_2 \le \cdots \le y_{m-1} \le n \\ \qquad x_i, y_j, \quad i = 1, 2, \cdots, n, \quad j = 1, 2, \cdots, m-1, \quad \text{integers.} \end{cases} \tag{10.11}$$

Chance Maximization Model

If we hope that total travel distance does not exceed a fixed number $\overline{g}$, then we have a distance constraint $g(\boldsymbol{x}, \boldsymbol{y}) \le \overline{g}$. If we want to maximize the chance that all customers are visited within their time windows subject to the distance constraint, then we have the following chance maximization model,

$$\begin{cases} \max \text{Ch}\{a_i \le f_i(\boldsymbol{x}, \boldsymbol{y}, \boldsymbol{t}) \le b_i, i = 1, 2, \cdots, n\} \\ \text{subject to:} \\ \qquad g(\boldsymbol{x}, \boldsymbol{y}) \le \overline{g} \\ \qquad 1 \le x_i \le n, \quad i = 1, 2, \cdots, n \\ \qquad x_i \ne x_j, \quad i \ne j, \ i, j = 1, 2, \cdots, n \\ \qquad 0 \le y_1 \le y_2 \le \cdots \le y_{m-1} \le n \\ \qquad x_i, y_j, \quad i = 1, 2, \cdots, n, \quad j = 1, 2, \cdots, m-1, \quad \text{integers.} \end{cases} \tag{10.12}$$

10.5 Exercises

Problem 10.1. Design a hybrid intelligent algorithm to solve hybrid models for vehicle routing problem (for example, the travel times are random and travel distances are fuzzy).

Problem 10.2. Build uncertain models for vehicle routing problem (for example, the travel times are uncertain variables with identification function (λ, ρ)), and design a hybrid intelligent algorithm to solve them.

Chapter 11
Facility Location Problem

Facility location problem is to find locations for new facilities such that the conveying cost from facilities to customers is minimized. Facility location problem has been studied for half a century because of its widely practical application backgrounds.

In practice, some factors such as demands, allocations, even locations of customers and facilities are usually changing. In an uncapacitated facility location problem, the customers are supplied by the nearest factory. However, in a capacitated problem, the customers may not be supplied by the nearest factory only. In order to solve this type of problem, this chapter introduces some optimization models for uncertain capacitated facility location problem.

11.1 Problem Description

In order to model facility location problem, we use the following indices, parameters, and decision variables:

$i = 1, 2, \cdots, n$: facilities;
$j = 1, 2, \cdots, m$: customers;
(a_j, b_j): location of customer j, $1 \le j \le m$;
ξ_j: uncertain demand of customer j, $1 \le j \le m$;
s_i: capacity of facility i, $1 \le i \le n$;
(x_i, y_i): decision vector representing the location of facility i, $1 \le i \le n$;
z_{ij}: quantity supplied to customer j by facility i after the uncertain demands are realized, $1 \le i \le n$, $1 \le j \le m$.

11.2 Stochastic Models

We write the demand vector $\boldsymbol{\xi} = (\xi_1, \xi_2, \cdots, \xi_m)$. For convenience, we also write

$$(\boldsymbol{x}, \boldsymbol{y}) = \begin{pmatrix} x_1 & y_1 \\ x_2 & y_2 \\ \cdots & \cdots \\ x_n & y_n \end{pmatrix}, \quad \boldsymbol{z} = \begin{pmatrix} z_{11} & z_{12} & \cdots & z_{1m} \\ z_{21} & z_{22} & \cdots & z_{2m} \\ \cdots & \cdots & \cdots & \cdots \\ z_{n1} & z_{n2} & \cdots & z_{nm} \end{pmatrix}. \tag{11.1}$$

B. Liu: Theory and Practice of Uncertain Programming, STUDFUZZ 239, pp. 157–165.
springerlink.com

For each $\omega \in \Omega$, $\boldsymbol{\xi}(\omega)$ is a realization of random vector $\boldsymbol{\xi}$. An allocation $\boldsymbol{z}$ is said to be feasible with respect to ω if and only if

$$\begin{cases} z_{ij} \geq 0, \ i = 1, 2, \cdots, n, \ j = 1, 2, \cdots, m \\ \sum_{i=1}^{n} z_{ij} = \xi_j(\omega), \ j = 1, 2, \cdots, m \\ \sum_{j=1}^{m} z_{ij} \leq s_i, \ i = 1, 2, \cdots, n. \end{cases} \tag{11.2}$$

We denote the feasible allocation set by

$$Z(\omega) = \left\{ \boldsymbol{z} \;\middle|\; \begin{array}{l} z_{ij} \geq 0, \ i = 1, 2, \cdots, n, \ j = 1, 2, \cdots, m \\ \sum_{i=1}^{n} z_{ij} = \xi_j(\omega), \ j = 1, 2, \cdots, m \\ \sum_{j=1}^{m} z_{ij} \leq s_i, \ i = 1, 2, \cdots, n \end{array} \right\}. \tag{11.3}$$

Note that $Z(\omega)$ may be an empty set for some ω.

For each $\omega \in \Omega$, the minimal cost is the one associated with the best allocation $\boldsymbol{z}$, i.e.,

$$C(\boldsymbol{x}, \boldsymbol{y}|\omega) = \min_{\boldsymbol{z} \in Z(\omega)} \sum_{i=1}^{n} \sum_{j=1}^{m} z_{ij} \sqrt{(x_i - a_j)^2 + (y_i - b_j)^2} \tag{11.4}$$

whose optimal solution $\boldsymbol{z}^*$ is called the optimal allocation. If $Z(\omega) = \emptyset$, then the demands of some customers are impossible to be met. As a penalty, we define

$$C(\boldsymbol{x}, \boldsymbol{y}|\omega) = \sum_{j=1}^{m} \max_{1 \leq i \leq n} \xi_j(\omega) \sqrt{(x_i - a_j)^2 + (y_i - b_j)^2}. \tag{11.5}$$

Stochastic Expected Cost Minimization Model

Since the demands are stochastic variables, the conveying cost $C(\boldsymbol{x}, \boldsymbol{y}|\omega)$ is also a stochastic variable. In order to evaluate the location design, we use its expected cost

$$E[C(\boldsymbol{x}, \boldsymbol{y}|\omega)] = \int_0^{\infty} \Pr\left\{\omega \in \boldsymbol{\Omega} \;\middle|\; C(\boldsymbol{x}, \boldsymbol{y}|\omega) \geq r\right\} \mathrm{d}r. \tag{11.6}$$

In order to minimize the expected cost, Zhou and Liu [337] presented the following expected cost minimization model for stochastic facility location problem,

$$\begin{cases} \min\limits_{\boldsymbol{x}, \boldsymbol{y}} \displaystyle\int_0^{\infty} \Pr\left\{\omega \in \boldsymbol{\Omega} \;\middle|\; C(\boldsymbol{x}, \boldsymbol{y}|\omega) \geq r\right\} \mathrm{d}r \\ \text{subject to:} \\ \qquad g_j(\boldsymbol{x}, \boldsymbol{y}) \leq 0, \ j = 1, 2, \cdots, p \end{cases} \tag{11.7}$$

where $g_j(\boldsymbol{x}, \boldsymbol{y}) \leq 0,\ j = 1, 2, \cdots, p$ represent the potential region of locations of new facilities and $C(\boldsymbol{x}, \boldsymbol{y}|\omega)$ is defined by (11.4).

This model is different from traditional stochastic programming models because there is a sub-problem in it, i.e.,

$$\begin{cases} \min \sum_{i=1}^{n} \sum_{j=1}^{m} z_{ij} \sqrt{(x_i - a_j)^2 + (y_i - b_j)^2} \\ \text{subject to:} \\ \qquad \sum_{i=1}^{n} z_{ij} = \xi_j(\omega), \quad j = 1, 2, \cdots, m \\ \qquad \sum_{j=1}^{m} z_{ij} \leq s_i, \quad i = 1, 2, \cdots, n \\ \qquad z_{ij} \geq 0, \quad i = 1, 2, \cdots, n, \quad j = 1, 2, \cdots, m. \end{cases} \tag{11.8}$$

Note that in (11.8) the parameters x_i, y_i and $\xi_j(\omega)$ are fixed real numbers for $i = 1, 2, \cdots, n$, $j = 1, 2, \cdots, m$. It is clearly a linear programming which may be solved by the simplex algorithm.

Example 11.1. Assume that there are 3 new facilities whose capacities are $(s_1, s_2, s_3) = (70, 80, 90)$, and 8 customers whose demands are uniformly distributed random variables. The locations (a_j, b_j) and demands $\mathcal{U}(l_i, u_i)$, $j = 1, 2, \cdots, 8$ of customers are given in Table 11.1.

Table 11.1 Locations and Random Demands of Customers

j	(a_j, b_j)	ξ_j	j	(a_j, b_j)	ξ_j
1	(28, 42)	(14,17)	5	(70, 18)	(21,26)
2	(18, 50)	(13,18)	6	(72, 98)	(24,28)
3	(74, 34)	(12,16)	7	(60, 50)	(13,16)
4	(74, 6)	(17,20)	8	(36, 40)	(12,17)

A run of the hybrid intelligent algorithm (5000 cycles in simulation, 300 generations in GA) shows that the optimal locations of the 3 facilities are

$$\begin{pmatrix} x_1^*, y_1^* \\ x_2^*, y_2^* \\ x_3^*, y_3^* \end{pmatrix} = \begin{pmatrix} 29.92, 43.19 \\ 70.04, 17.99 \\ 72.02, 98.02 \end{pmatrix}$$

whose expected conveying cost is 1259.

Stochastic α-Cost Minimization Model

Now we define the α-cost as the minimum number $\overline{C}$ such that

$$\Pr\{C(\boldsymbol{x}, \boldsymbol{y}|\omega) \leq \overline{C}\} \geq \alpha. \tag{11.9}$$

If we want to minimize the α-cost, then we have the following stochastic α-cost minimization model (Zhou and Liu [337]),

$$\begin{cases} \min_{\boldsymbol{x},\boldsymbol{y}} \overline{C} \\ \text{subject to:} \\ \qquad \Pr\left\{\omega \in \Omega \mid C(\boldsymbol{x},\boldsymbol{y}|\omega) \le \overline{C}\right\} \ge \alpha \\ \qquad g_j(\boldsymbol{x},\boldsymbol{y}) \le 0, \ j = 1, 2, \cdots, p \end{cases} \tag{11.10}$$

where $\overline{f}$ is the α-cost and $C(\boldsymbol{x},\boldsymbol{y}|\omega)$ is defined by (11.4).

Example 11.2. Here we suppose that the 0.9-cost is to be minimized. A run of the hybrid intelligent algorithm (5000 cycles in simulations and 300 generations in GA) shows that the optimal locations of the 3 facilities are

$$\begin{pmatrix} x_1^*, y_1^* \\ x_2^*, y_2^* \\ x_3^*, y_3^* \end{pmatrix} = \begin{pmatrix} 31.00, 43.08 \\ 70.04, 17.92 \\ 71.98, 97.99 \end{pmatrix}$$

whose 0.9-cost is 1313.

Probability Maximization Model

If we hope to maximize the probability that the conveying cost will not exceed a given level C^0, then we have a probability maximization model (Zhou and Liu [337]),

$$\begin{cases} \max_{\boldsymbol{x},\boldsymbol{y}} \Pr\left\{\omega \in \Omega \mid C(\boldsymbol{x},\boldsymbol{y}|\omega) \le C^0\right\} \\ \text{subject to:} \\ \qquad g_j(\boldsymbol{x},\boldsymbol{y}) \le 0, \ j = 1, 2, \cdots, p \end{cases} \tag{11.11}$$

where $C(\boldsymbol{x},\boldsymbol{y}|\omega)$ is defined by (11.4).

Example 11.3. Now we want to maximize the probability that the transportation cost does not exceed 1300. A run of the hybrid intelligent algorithm (5000 cycles in simulation, and 300 generations in GA) shows that the optimal locations of the 3 facilities are

$$\begin{pmatrix} x_1^*, y_1^* \\ x_2^*, y_2^* \\ x_3^*, y_3^* \end{pmatrix} = \begin{pmatrix} 30.21, 42.75 \\ 70.47, 17.94 \\ 72.03, 97.98 \end{pmatrix}$$

whose probability is 0.86.

11.3 Fuzzy Models

In this section, expert knowledge is used to estimate the demands. Thus we have facility location problem with fuzzy demands.

Fuzzy Expected Cost Minimization Model

In this section, we suppose that ξ_j are fuzzy demands of customers j defined on the credibility space $(\Theta, \mathcal{P}, \text{Cr})$, $j = 1, 2, \cdots, m$, respectively, and denote $\boldsymbol{\xi} = (\xi_1, \xi_2, \cdots, \xi_n)$.

For each $\theta \in \Theta$, $\boldsymbol{\xi}(\theta)$ is a realization of fuzzy vector $\boldsymbol{\xi}$. We denote the feasible allocation set by

$$Z(\theta) = \left\{ \boldsymbol{z} \;\middle|\; \begin{array}{l} z_{ij} \geq 0, \; i = 1, 2, \cdots, n, \; j = 1, 2, \cdots, m \\ \sum_{i=1}^{n} z_{ij} = \xi_j(\theta), \; j = 1, 2, \cdots, m \\ \sum_{j=1}^{m} z_{ij} \leq s_i, \; i = 1, 2, \cdots, n \end{array} \right\}. \tag{11.12}$$

Note that $Z(\theta)$ may be an empty set for some θ.

For each $\theta \in \Theta$, the minimal conveying cost from facilities to customers is

$$C(\boldsymbol{x}, \boldsymbol{y}|\theta) = \min_{\boldsymbol{z} \in Z(\theta)} \sum_{i=1}^{n} \sum_{j=1}^{m} z_{ij} \sqrt{(x_i - a_j)^2 + (y_i - b_j)^2} \tag{11.13}$$

whose optimal solution $\boldsymbol{z}^*$ is called the optimal allocation. If $Z(\theta) = \emptyset$, then the demands of some customers are impossible to be met. As a penalty, we define

$$C(\boldsymbol{x}, \boldsymbol{y}|\theta) = \sum_{j=1}^{m} \max_{1 \leq i \leq n} \xi_j(\theta) \sqrt{(x_i - a_j)^2 + (y_i - b_j)^2}. \tag{11.14}$$

Note that the conveying cost $C(\boldsymbol{x}, \boldsymbol{y}|\theta)$ is a fuzzy variable. In order to evaluate the location design, we use its expected cost

$$E[C(\boldsymbol{x}, \boldsymbol{y}|\theta)] = \int_0^{\infty} \text{Cr}\left\{\theta \in \boldsymbol{\Theta} \;\middle|\; C(\boldsymbol{x}, \boldsymbol{y}|\theta) \geq r\right\} \mathrm{d}r. \tag{11.15}$$

In order to minimize the expected cost, Zhou and Liu [339] presented the following expected cost minimization model for fuzzy capacitated facility location problem,

$$\left\{ \begin{array}{l} \min\limits_{\boldsymbol{x}, \boldsymbol{y}} \displaystyle\int_0^{\infty} \text{Cr}\left\{\theta \in \Theta | C(\boldsymbol{x}, \boldsymbol{y}|\theta) \geq r\right\} \mathrm{d}r \\ \text{subject to:} \\ \qquad g_j(\boldsymbol{x}, \boldsymbol{y}) \leq 0, \; j = 1, 2, \cdots, p \end{array} \right. \tag{11.16}$$

where $g_j(\boldsymbol{x}, \boldsymbol{y}) \leq 0$, $j = 1, 2, \cdots, p$ represent the potential region of locations of new facilities and $C(\boldsymbol{x}, \boldsymbol{y}|\theta)$ is defined by (11.13).

Example 11.4. Now we assume that there are 8 customers whose locations and trapezoidal fuzzy demands are given in Table 11.2, and 3 facilities with capacities 70, 80 and 90.

Table 11.2 Locations and Fuzzy Demands of Customers

j	(a_j, b_j)	ξ_j	j	(a_j, b_j)	ξ_j
1	(28, 42)	(14,15,16,17)	5	(70, 18)	(21,23,24,26)
2	(18, 50)	(13,14,16,18)	6	(72, 98)	(24,25,26,28)
3	(74, 34)	(12,14,15,16)	7	(60, 50)	(13,14,15,16)
4	(74, 6)	(17,18,19,20)	8	(36, 40)	(12,14,16,17)

A run of the hybrid intelligent algorithm (10000 cycles in fuzzy simulation, 1000 generations in GA) shows that the optimal locations of the 3 facilities are

$$\begin{pmatrix} x_1^*, y_1^* \\ x_2^*, y_2^* \\ x_3^*, y_3^* \end{pmatrix} = \begin{pmatrix} 30.34, 42.50 \\ 70.14, 17.97 \\ 71.99, 98.04 \end{pmatrix}$$

whose expected conveying cost is 1255.

Fuzzy α-Cost Minimization Model

Now we define the α-cost as the minimum number $\overline{f}$ such that

$$\mathrm{Cr}\{C(\boldsymbol{x}, \boldsymbol{y}|\theta) \le \overline{f}\} \ge \alpha.$$

If we want to minimize the α-cost, then we have a fuzzy α-cost minimization model (Zhou and Liu [339]),

$$\begin{cases} \min\limits_{\boldsymbol{x},\boldsymbol{y}} \overline{f} \\ \text{subject to:} \\ \qquad \mathrm{Cr}\left\{\theta \in \Theta \;\middle|\; C(\boldsymbol{x}, \boldsymbol{y}|\theta) \le \overline{f}\right\} \ge \alpha \\ \qquad g_j(\boldsymbol{x}, \boldsymbol{y}) \le 0, \; j = 1, 2, \cdots, p \end{cases} \tag{11.17}$$

where $\overline{f}$ is the α-cost and $C(\boldsymbol{x}, \boldsymbol{y}|\theta)$ is defined by (11.13).

Example 11.5. Let us minimize the 0.9-cost. A run of the hybrid intèlligent algorithm (10000 cycles in fuzzy simulation, 1000 generations in GA) shows that the optimal locations of the 3 facilities are

$$\begin{pmatrix} x_1^*, y_1^* \\ x_2^*, y_2^* \\ x_3^*, y_3^* \end{pmatrix} = \begin{pmatrix} 30.39, 43.49 \\ 70.10, 17.64 \\ 71.98, 98.03 \end{pmatrix}$$

whose 0.9-cost is 1354.

Credibility Maximization Model

If we hope to maximize the credibility that the conveying cost does not exceed a given level C^0, then we have a credibility maximization model (Zhou and Liu [339]),

$$\begin{cases} \max\limits_{\boldsymbol{x},\boldsymbol{y}} \ \mathrm{Cr}\left\{\theta \in \Theta \mid C(\boldsymbol{x},\boldsymbol{y}|\theta) \le C^0\right\} \\ \text{subject to:} \\ \qquad g_j(\boldsymbol{x},\boldsymbol{y}) \le 0, \quad j = 1,2,\cdots,p \end{cases} \tag{11.18}$$

where $C(\boldsymbol{x},\boldsymbol{y}|\theta)$ is defined by (11.13).

Example 11.6. Assume $C^0 = 1350$. A run of the hybrid intelligent algorithm (10000 cycles in fuzzy simulation, 1000 generations in GA) shows that the optimal locations of the 3 facilities are

$$\begin{pmatrix} x_1^*, y_1^* \\ x_2^*, y_2^* \\ x_3^*, y_3^* \end{pmatrix} = \begin{pmatrix} 32.18, 42.04 \\ 70.15, 17.81 \\ 72.04, 98.09 \end{pmatrix}$$

whose credibility is 0.87.

11.4 Hybrid Models

In this section, we suppose that the demands of customers are hybrid variables defined on $(\Theta, \mathcal{P}, \mathrm{Cr}) \times (\Omega, \mathcal{A}, \Pr)$.

For each $(\theta,\omega) \in \Theta \times \Omega$, the value $\boldsymbol{\xi}(\theta,\omega)$ is a realization of hybrid vector $\boldsymbol{\xi}$. We denote the feasible allocation set by

$$Z(\theta,\omega) = \left\{ \boldsymbol{z} \ \middle| \ \begin{array}{l} z_{ij} \ge 0, \ i = 1,2,\cdots,n, \ j = 1,2,\cdots,m \\ \sum\limits_{i=1}^{n} z_{ij} = \xi_j(\theta,\omega), \ j = 1,2,\cdots,m \\ \sum\limits_{j=1}^{m} z_{ij} \le s_i, \ i = 1,2,\cdots,n \end{array} \right\}. \tag{11.19}$$

Note that $Z(\theta,\omega)$ may be an empty set for some (θ,ω).

For each $(\theta,\omega) \in \Theta \times \Omega$, the minimal conveying cost from facilities to customers is

$$C(\boldsymbol{x},\boldsymbol{y}|\theta,\omega) = \min_{\boldsymbol{z}\in Z(\theta,\omega)} \sum_{i=1}^{n}\sum_{j=1}^{m} z_{ij}\sqrt{(x_i-a_j)^2+(y_i-b_j)^2} \tag{11.20}$$

whose optimal solution $\boldsymbol{z}^*$ is called the optimal allocation. If $Z(\theta,\omega) = \emptyset$, then the demands of some customers are impossible to be met. As a penalty, we define

$$C(\boldsymbol{x},\boldsymbol{y}|\theta,\omega) = \sum_{j=1}^{m} \max_{1\le i\le n} \xi_j(\theta,\omega)\sqrt{(x_i-a_j)^2+(y_i-b_j)^2}. \tag{11.21}$$

Note that the conveying cost $C(\boldsymbol{x},\boldsymbol{y}|\theta,\omega)$ is a hybrid variable. In order to evaluate the location design, we use its expected cost

$$E[C(\boldsymbol{x},\boldsymbol{y}|\theta,\omega)] = \int_0^{\infty} \mathrm{Ch}\left\{(\theta,\omega)\in\Theta\times\Omega \;\middle|\; C(\boldsymbol{x},\boldsymbol{y}|\theta,\omega)\ge r\right\}\mathrm{d}r.$$

Hybrid Expected Cost Minimization Model

In order to minimize the expected cost, we have the following expected cost minimization model for hybrid capacitated facility location problem,

$$\begin{cases}\displaystyle\min_{\boldsymbol{x},\boldsymbol{y}} \int_0^{\infty} \mathrm{Ch}\left\{(\theta,\omega)\in\Theta\times\Omega|C(\boldsymbol{x},\boldsymbol{y}|\theta,\omega)\ge r\right\}\mathrm{d}r\\ \text{subject to:}\\ \qquad g_j(\boldsymbol{x},\boldsymbol{y})\le 0,\ \ j=1,2,\cdots,p\end{cases} \tag{11.22}$$

where $g_j(\boldsymbol{x},\boldsymbol{y})\le 0$, $j=1,2,\cdots,p$ represent the potential region of locations of new facilities and $C(\boldsymbol{x},\boldsymbol{y}|\theta,\omega)$ is defined by (11.20).

Hybrid α-Cost Minimization Model

Now we define the α-cost as the minimum number $\overline{f}$ such that

$$\mathrm{Ch}\{C(\boldsymbol{x},\boldsymbol{y}|\theta,\omega)\le\overline{f}\}\ge\alpha.$$

If we want to minimize the α-cost, then we have a hybrid α-cost minimization model,

$$\begin{cases}\displaystyle\min_{\boldsymbol{x},\boldsymbol{y}}\ \overline{f}\\ \text{subject to:}\\ \qquad \mathrm{Ch}\left\{(\theta,\omega)\in\Theta\times\Omega \;\middle|\; C(\boldsymbol{x},\boldsymbol{y}|\theta,\omega)\le\overline{f}\right\}\ge\alpha\\ \qquad g_j(\boldsymbol{x},\boldsymbol{y})\le 0,\ \ j=1,2,\cdots,p\end{cases} \tag{11.23}$$

where $\overline{f}$ is the α-cost and $C(\boldsymbol{x},\boldsymbol{y}|\theta,\omega)$ is defined by (11.20).

Chance Maximization Model

If we hope to maximize the chance that the conveying cost does not exceed a given level C^0, then we have a chance maximization model,

$$\begin{cases}\displaystyle\max_{\boldsymbol{x},\boldsymbol{y}}\ \mathrm{Ch}\left\{(\theta,\omega)\in\Theta\times\Omega \;\middle|\; C(\boldsymbol{x},\boldsymbol{y}|\theta,\omega)\le C^0\right\}\\ \text{subject to:}\\ \qquad g_j(\boldsymbol{x},\boldsymbol{y})\le 0,\ \ j=1,2,\cdots,p\end{cases} \tag{11.24}$$

where $C(\boldsymbol{x},\boldsymbol{y}|\theta,\omega)$ is defined by (11.20).

11.5 Exercises

Problem 11.1. Design a hybrid intelligent algorithm to solve hybrid models for facility location problem (for example, the demands of customers are random and the locations of customers are fuzzy).

Problem 11.2. Build uncertain models for facility location problem (for example, the demands of customers are uncertain variables with identification function (λ, ρ)), and design a hybrid intelligent algorithm to solve them.

Chapter 12
Machine Scheduling Problem

Machine scheduling problem is concerned with finding an efficient schedule during an uninterrupted period of time for a set of machines to process a set of jobs. Much of research work has been done on this type of problem during the past five decades.

12.1 Problem Description

In a machine scheduling problem, we assume that (a) each job can be processed on any machine without interruption; and (b) each machine can process only one job at a time.

Let us first introduce the following indices and parameters.

$i = 1, 2, \cdots, n$: jobs;

$k = 1, 2, \cdots, m$: machines;

ξ_{ik}: uncertain processing time of job i on machine k;

D_i: the due date of job i, $i = 1, 2, \cdots, n$.

The schedule is represented by Liu's formulation [181] via two decision vectors $\boldsymbol{x}$ and $\boldsymbol{y}$, where

$\boldsymbol{x} = (x_1, x_2, \cdots, x_n)$: integer decision vector representing n jobs with $1 \le x_i \le n$ and $x_i \ne x_j$ for all $i \ne j$, $i, j = 1, 2, \cdots, n$. That is, the sequence $\{x_1, x_2, \cdots, x_n\}$ is a rearrangement of $\{1, 2, \cdots, n\}$;

$\boldsymbol{y} = (y_1, y_2, \cdots, y_{m-1})$: integer decision vector with $y_0 \equiv 0 \le y_1 \le y_2 \le \cdots \le y_{m-1} \le n \equiv y_m$.

We note that the schedule is fully determined by the decision vectors $\boldsymbol{x}$ and $\boldsymbol{y}$ in the following way. For each k $(1 \le k \le m)$, if $y_k = y_{k-1}$, then machine k is not used; if $y_k > y_{k-1}$, then machine k is used and processes jobs $x_{y_{k-1}+1}, x_{y_{k-1}+2}, \cdots, x_{y_k}$ in turn. Thus the schedule of all machines is as follows:

$$
\begin{aligned}
&\text{Machine 1: } x_{y_0+1} \to x_{y_0+2} \to \cdots \to x_{y_1};\\
&\text{Machine 2: } x_{y_1+1} \to x_{y_1+2} \to \cdots \to x_{y_2};\\
&\quad \cdots\\
&\text{Machine } m\text{: } x_{y_{m-1}+1} \to x_{y_{m-1}+2} \to \cdots \to x_{y_m}.
\end{aligned}
\tag{12.1}
$$

Let $C_i(\boldsymbol{x}, \boldsymbol{y}, \boldsymbol{\xi})$ be the completion times of jobs i, $i = 1, 2, \cdots, n$, respectively. They can be calculated by the following equations,

B. Liu: Theory and Practice of Uncertain Programming, STUDFUZZ 239, pp. 167–177.
springerlink.com

$$C_{x_{y_{k-1}+1}}(\boldsymbol{x}, \boldsymbol{y}, \boldsymbol{\xi}) = \xi_{x_{y_{k-1}+1}k} \tag{12.2}$$

and

$$C_{x_{y_{k-1}+j}}(\boldsymbol{x}, \boldsymbol{y}, \boldsymbol{\xi}) = C_{x_{y_{k-1}+j-1}}(\boldsymbol{x}, \boldsymbol{y}, \boldsymbol{\xi}) + \xi_{x_{y_{k-1}+j}k} \tag{12.3}$$

for $2 \le j \le y_k - y_{k-1}$ and $k = 1, 2, \cdots, m$.

We denote the tardiness and makespan of the schedule $(\boldsymbol{x}, \boldsymbol{y})$ by $f_1(\boldsymbol{x}, \boldsymbol{y}, \boldsymbol{\xi})$ and $f_2(\boldsymbol{x}, \boldsymbol{y}, \boldsymbol{\xi})$, respectively. Then we have

$$f_1(\boldsymbol{x}, \boldsymbol{y}, \boldsymbol{\xi}) = \max_{1 \le i \le n} \{C_i(\boldsymbol{x}, \boldsymbol{y}, \boldsymbol{\xi}) - D_i\} \vee 0, \tag{12.4}$$

$$f_2(\boldsymbol{x}, \boldsymbol{y}, \boldsymbol{\xi}) = \max_{1 \le k \le m} C_{x_{y_k}}(\boldsymbol{x}, \boldsymbol{y}, \boldsymbol{\xi}). \tag{12.5}$$

12.2 Stochastic Models

In this section, we introduce an expected time goal programming model for parallel machine scheduling problems proposed by Peng and Liu [254].

Stochastic Expected Time Goal Programming

In order to balance the above conflicting objectives, we may have the following target levels and priority structure.

At the first priority level, the expected tardiness $E[f_1(\boldsymbol{x}, \boldsymbol{y}, \boldsymbol{\xi})]$ should not exceed the target value b_1. Thus we have a goal constraint

$$E[f_1(\boldsymbol{x}, \boldsymbol{y}, \boldsymbol{\xi})] - b_1 = d_1^+$$

in which $d_1^+ \vee 0$ will be minimized.

At the second priority level, the expected makespan $E[f_2(\boldsymbol{x}, \boldsymbol{y}, \boldsymbol{\xi})]$ should not exceed the target value b_2. That is, we have a goal constraint

$$E[f_2(\boldsymbol{x}, \boldsymbol{y}, \boldsymbol{\xi})] - b_2 = d_2^+$$

in which $d_2^+ \vee 0$ will be minimized.

Then according to the priority structure, we have the following stochastic expected time goal programming model:

$$\begin{cases} \text{lexmin} \left\{d_1^+ \vee 0, d_2^+ \vee 0\right\} \\ \text{subject to:} \\ \quad E[f_1(\boldsymbol{x}, \boldsymbol{y}, \boldsymbol{\xi})] - b_1 = d_1^+ \\ \quad E[f_2(\boldsymbol{x}, \boldsymbol{y}, \boldsymbol{\xi})] - b_2 = d_2^+ \\ \quad 1 \le x_i \le n, \quad i = 1, 2, \cdots, n \\ \quad x_i \ne x_j, \quad i \ne j, \ i, j = 1, 2, \cdots, n \\ \quad 0 \le y_1 \le y_2 \cdots \le y_{m-1} \le n \\ \quad x_i, y_j, \quad i = 1, 2, \cdots, n, \quad j = 1, 2, \cdots, m-1, \quad \text{integers.} \end{cases} \tag{12.6}$$

Hybrid Intelligent Algorithm

The hybrid intelligent algorithm documented in Chapter 4 may solve this model provided that the initialization, crossover and mutation operations are revised as follows.

We encode a schedule into a chromosome $V = (\boldsymbol{x}, \boldsymbol{y})$, where $\boldsymbol{x}, \boldsymbol{y}$ are the same as the decision vectors. For the gene section $\boldsymbol{x}$, we define a sequence $\{x_1, x_2, \cdots, x_n\}$ with $x_i = i$, $i = 1, 2, \cdots, n$. In order to get a random rearrangement of $\{1, 2, \cdots, n\}$, we repeat the following process from $j = 1$ to n: generating a random position n' between j and n, and exchanging the values of x_j and $x_{n'}$. For each i with $1 \le i \le m-1$, we set y_i as a random integer between 0 and n. Then we rearrange the sequence $\{y_1, y_2, \cdots, y_{m-1}\}$ from small to large and thus obtain a gene section $\boldsymbol{y} = (y_1, y_2, \cdots, y_{m-1})$. We can ensure that the produced chromosome $V = (\boldsymbol{x}, \boldsymbol{y})$ is always feasible.

Let us illustrate the crossover operator on the pair V_1 and V_2. We denote $V_1 = (\boldsymbol{x}_1, \boldsymbol{y}_1)$ and $V_2 = (\boldsymbol{x}_2, \boldsymbol{y}_2)$. Two children V_1' and V_2' are produced by the crossover operation as follows: $V_1' = (\boldsymbol{x}_1, \boldsymbol{y}_2)$ and $V_2' = (\boldsymbol{x}_2, \boldsymbol{y}_1)$. Note that the obtained chromosomes $V_1' = (\boldsymbol{x}_1, \boldsymbol{y}_2)$ and $V_2' = (\boldsymbol{x}_2, \boldsymbol{y}_1)$ in this way are always feasible.

We mutate the parent $V = (\boldsymbol{x}, \boldsymbol{y})$ in the following way. For the gene $\boldsymbol{x}$, we randomly generate two mutation positions n_1 and n_2 between 1 and n, and rearrange the sequence $\{x_{n_1}, x_{n_1+1}, \cdots, x_{n_2}\}$ at random to form a new sequence $\{x'_{n_1}, x'_{n_1+1}, \cdots, x'_{n_2}\}$, thus we obtain a new gene section

$$\boldsymbol{x}' = (x_1, \cdots, x_{n_1-1}, x'_{n_1}, x'_{n_1+1}, \cdots, x'_{n_2}, x_{n_2+1}, \cdots, x_n).$$

Similarly, for the gene $\boldsymbol{y}$, we generate two random mutation positions n_1 and n_2 between 1 and $m-1$, and set y_i as a random integer number y_i' between 0 and n for $i = n_1, n_1+1, \cdots, n_2$. We then rearrange the sequence $y_1, \cdots, y_{n_1-1}, y'_{n_1}, y'_{n_1+1}, \cdots, y'_{n_2}, y_{n_2+1}, \cdots, y_{m-1}$ from small to large and obtain a new gene section $\boldsymbol{y}'$. Finally, we replace the parent V with the offspring $V' = (\boldsymbol{x}', \boldsymbol{y}')$.

Example 12.1. Assume that there are 8 jobs and 3 machines. The processing times of jobs on different machines are all uniformly distributed random variables. The random processing times and due dates are listed in Table 12.1.

We suppose that the target levels of expected tardiness and expected makespan are $b_1 = 0$ and $b_2 = 43$. A run of the hybrid intelligent algorithm (10000 cycles in simulation, 1000 generations in GA) shows that the optimal schedule is

$$\begin{aligned} &\text{Machine 1:} \quad 1 \to 5 \to 3; \\ &\text{Machine 2:} \quad 6 \to 7 \to 2; \\ &\text{Machine 3:} \quad 8 \to 4 \end{aligned}$$

which can satisfy the two goals.

Table 12.1 Random Processing Times and Due Dates

Jobs	Random Processing Times			Due Dates
	Machine 1	Machine 2	Machine 3	
1	$(10, 16)$	$(11, 15)$	$(12, 17)$	30
2	$(10, 18)$	$(10, 16)$	$(12, 18)$	150
3	$(12, 16)$	$(11, 16)$	$(12, 18)$	105
4	$(18, 24)$	$(20, 23)$	$(20, 25)$	130
5	$(10, 15)$	$(12, 17)$	$(10, 15)$	60
6	$(10, 18)$	$(12, 17)$	$(12, 20)$	30
7	$(10, 16)$	$(10, 14)$	$(10, 13)$	75
8	$(15, 20)$	$(14, 20)$	$(12, 16)$	50

Stochastic Chance-Constrained Goal Programming

We assume the following priority structure. At the first priority level, the tardiness $f_1(\boldsymbol{x}, \boldsymbol{y}, \boldsymbol{\xi})$ should not exceed the target value b_1 with a confidence level α_1. Thus we have a goal constraint

$$\Pr\left\{f_1(\boldsymbol{x}, \boldsymbol{y}, \boldsymbol{\xi}) - b_1 \le d_1^+\right\} \ge \alpha_1$$

in which $d_1^+ \vee 0$ will be minimized.

At the second priority level, the makespan $f_2(\boldsymbol{x}, \boldsymbol{y}, \boldsymbol{\xi})$ should not exceed the target value b_2 with a confidence level α_2. Thus we have a goal constraint

$$\Pr\left\{f_2(\boldsymbol{x}, \boldsymbol{y}, \boldsymbol{\xi}) - b_2 \le d_2^+\right\} \ge \alpha_2$$

in which $d_2^+ \vee 0$ will be minimized.

Then we have the following CCGP model for parallel machine scheduling problem (Peng and Liu [254]),

$$\left\{\begin{array}{l}
\text{lexmin}\{d_1^+ \vee 0, d_2^+ \vee 0\} \\
\text{subject to:} \\
\quad \Pr\left\{f_1(\boldsymbol{x}, \boldsymbol{y}, \boldsymbol{\xi}) - b_1 \le d_1^+\right\} \ge \alpha_1 \\
\quad \Pr\left\{f_2(\boldsymbol{x}, \boldsymbol{y}, \boldsymbol{\xi}) - b_2 \le d_2^+\right\} \ge \alpha_2 \\
\quad 1 \le x_i \le n, \quad i = 1, 2, \cdots, n \\
\quad x_i \ne x_j, \quad i \ne j, \ i, j = 1, 2, \cdots, n \\
\quad 0 \le y_1 \le y_2 \cdots \le y_{m-1} \le n \\
\quad x_i, \ y_j, \quad i = 1, 2, \cdots, n, \quad j = 1, 2, \cdots, m-1, \quad \text{integers.}
\end{array}\right. \tag{12.7}$$

Example 12.2. Suppose that the target levels of tardiness and makespan are $b_1 = 0$ and $b_2 = 43$ with confidence levels $\alpha_1 = 0.95$ and $\alpha_2 = 0.90$. A run of the hybrid intelligent algorithm (10000 cycles in simulation, 1000 generations in GA) shows that the optimal schedule is

Machine 1: $5 \to 1 \to 3$;
Machine 2: $7 \to 6 \to 2$;
Machine 3: $8 \to 4$

which can satisfy the first goal, but the second objective is 0.52.

Stochastic Dependent-Chance Goal Programming

Suppose that the management goals have the following priority structure.

At the first priority level, the probability that the tardiness $f_1(\boldsymbol{x}, \boldsymbol{y}, \boldsymbol{\xi})$ does not exceed the given value b_1 should achieve a confidence level α_1. Thus we have a goal constraint

$$\alpha_1 - \Pr\{f_1(\boldsymbol{x}, \boldsymbol{y}, \boldsymbol{\xi}) \le b_1\} = d_1^-$$

in which $d_1^- \vee 0$ will be minimized.

At the second priority level, the probability that the makespan $f_2(\boldsymbol{x}, \boldsymbol{y}, \boldsymbol{\xi})$ does not exceed the given value b_2 should achieve a confidence level α_2. Thus we have a goal constraint

$$\alpha_2 - \Pr\{f_2(\boldsymbol{x}, \boldsymbol{y}, \boldsymbol{\xi}) \le b_2\} = d_2^-$$

in which $d_2^- \vee 0$ will be minimized.

Then we have the following DCGP model formulated by Peng and Liu [254],

$$\begin{cases} \text{lexmin}\{d_1^- \vee 0, d_2^- \vee 0\} \\ \text{subject to:} \\ \quad \alpha_1 - \Pr\{f_1(\boldsymbol{x}, \boldsymbol{y}, \boldsymbol{\xi}) \le b_1\} = d_1^- \\ \quad \alpha_2 - \Pr\{f_2(\boldsymbol{x}, \boldsymbol{y}, \boldsymbol{\xi}) \le b_2\} = d_2^- \\ \quad 1 \le x_i \le n, \quad i = 1, 2, \cdots, n \\ \quad x_i \ne x_j, \quad i \ne j, \ i, j = 1, 2, \cdots, n \\ \quad 0 \le y_1 \le y_2 \cdots \le y_{m-1} \le n \\ \quad x_i, y_j, \quad i = 1, 2, \cdots, n, \quad j = 1, 2, \cdots, m-1, \quad \text{integers.} \end{cases} \tag{12.8}$$

Example 12.3. Suppose that the upper bounds of tardiness and makespan are $b_1 = 0$ and $b_2 = 43$, and the target levels are $\alpha_1 = 0.95$ and $\alpha_2 = 0.90$. A run of the hybrid intelligent algorithm (10000 cycles in simulation, 1000 generations in GA) shows that the optimal schedule is

Machine 1: $1 \to 5 \to 3$;
Machine 2: $6 \to 2 \to 7$;
Machine 3: $8 \to 4$

which can satisfy the first goal, but the second objective is 0.05.

12.3 Fuzzy Models

In this section, we assume that the processing times are fuzzy variables, and construct fuzzy models for machine scheduling problems.

Fuzzy Expected Time Goal Programming

In order to balance the conflicting objectives, we may have the following target levels and priority structure:

At the first priority level, the expected tardiness $E[f_1(\boldsymbol{x},\boldsymbol{y},\boldsymbol{\xi})]$ should not exceed the target value b_1. Thus we have a goal constraint

$$E[f_1(\boldsymbol{x},\boldsymbol{y},\boldsymbol{\xi})] - b_1 = d_1^+$$

in which $d_1^+ \vee 0$ will be minimized.

At the second priority level, the expected makespan $E[f_2(\boldsymbol{x},\boldsymbol{y},\boldsymbol{\xi})]$ should not exceed the target value b_2. That is, we have a goal constraint

$$E[f_2(\boldsymbol{x},\boldsymbol{y},\boldsymbol{\xi})] - b_2 = d_2^+$$

in which $d_2^+ \vee 0$ will be minimized.

Then we have the following fuzzy expected time goal programming model for the parallel machine scheduling problem (Peng and Liu [255]),

$$\begin{cases} \text{lexmin}\left\{d_1^+ \vee 0, d_2^+ \vee 0\right\} \\ \text{subject to:} \\ \quad E[f_1(\boldsymbol{x},\boldsymbol{y},\boldsymbol{\xi})] - b_1 = d_1^+ \\ \quad E[f_2(\boldsymbol{x},\boldsymbol{y},\boldsymbol{\xi})] - b_2 = d_2^+ \\ \quad 1 \le x_i \le n, \quad i = 1,2,\cdots,n \\ \quad x_i \ne x_j, \quad i \ne j, \; i,j = 1,2,\cdots,n \\ \quad 0 \le y_1 \le y_2 \cdots \le y_{m-1} \le n \\ \quad x_i, y_j, \quad i = 1,2,\cdots,n, \quad j = 1,2,\cdots,m-1, \quad \text{integers.} \end{cases} \tag{12.9}$$

Example 12.4. Assume that there are 8 jobs and 3 machines. The trapezoidal fuzzy processing times and the due dates are listed in Table 12.2.

Suppose that $b_1 = 0$ and $b_2 = 45$. A run of the hybrid intelligent algorithm (10000 cycles in fuzzy simulation, 500 generations in GA) shows that the optimal schedule is

$$\begin{aligned} &\text{Machine 1:} \quad 1 \to 3 \to 5; \\ &\text{Machine 2:} \quad 6 \to 7 \to 2; \\ &\text{Machine 3:} \quad 4 \to 8 \end{aligned}$$

which can satisfy the two goals.

Table 12.2 Fuzzy Processing Times and Due Dates

Jobs	Fuzzy Processing Times			Due Dates
	Machine 1	Machine 2	Machine 3	
1	$(10, 11, 14, 16)$	$(11, 12, 14, 15)$	$(12, 13, 15, 17)$	30
2	$(10, 13, 16, 18)$	$(10, 11, 15, 16)$	$(12, 13, 14, 18)$	150
3	$(12, 14, 15, 16)$	$(11, 13, 14, 16)$	$(12, 15, 16, 18)$	105
4	$(18, 21, 22, 24)$	$(20, 21, 22, 23)$	$(20, 23, 24, 25)$	130
5	$(10, 12, 13, 15)$	$(12, 14, 15, 17)$	$(10, 12, 13, 15)$	60
6	$(10, 14, 15, 18)$	$(12, 13, 16, 17)$	$(12, 16, 17, 20)$	30
7	$(10, 11, 12, 16)$	$(10, 11, 12, 14)$	$(10, 11, 12, 13)$	75
8	$(15, 17, 18, 20)$	$(14, 15, 16, 20)$	$(12, 14, 15, 16)$	50

Fuzzy Chance-Constrained Goal Programming

We assume the following priority structure. At the first priority level, the tardiness $f_1(\boldsymbol{x}, \boldsymbol{y}, \boldsymbol{\xi})$ should not exceed the target value b_1 with a confidence level α_1. Thus we have a goal constraint

$$\text{Cr}\left\{f_1(\boldsymbol{x}, \boldsymbol{y}, \boldsymbol{\xi}) - b_1 \le d_1^+\right\} \ge \alpha_1$$

in which $d_1^+ \vee 0$ will be minimized.

At the second priority level, the makespan $f_2(\boldsymbol{x}, \boldsymbol{y}, \boldsymbol{\xi})$ should not exceed the target value b_2 with a confidence level α_2. Thus we have a goal constraint

$$\text{Cr}\left\{f_2(\boldsymbol{x}, \boldsymbol{y}, \boldsymbol{\xi}) - b_2 \le d_2^+\right\} \ge \alpha_2$$

in which $d_2^+ \vee 0$ will be minimized.

Then we have the following fuzzy CCGP model for the parallel machine scheduling problem (Peng and Liu [255]),

$$\begin{cases} \text{lexmin}\{d_1^+ \vee 0, d_2^+ \vee 0\} \\ \text{subject to:} \\ \qquad \text{Cr}\left\{f_1(\boldsymbol{x}, \boldsymbol{y}, \boldsymbol{\xi}) - b_1 \le d_1^+\right\} \ge \alpha_1 \\ \qquad \text{Cr}\left\{f_2(\boldsymbol{x}, \boldsymbol{y}, \boldsymbol{\xi}) - b_2 \le d_2^+\right\} \ge \alpha_2 \\ \qquad 1 \le x_i \le n, \quad i = 1, 2, \cdots, n \\ \qquad x_i \ne x_j, \quad i \ne j, \ i, j = 1, 2, \cdots, n \\ \qquad 0 \le y_1 \le y_2 \cdots \le y_{m-1} \le n \\ \qquad x_i, \ y_j, \quad i = 1, 2, \cdots, n, \quad j = 1, 2, \cdots, m-1, \quad \text{integers.} \end{cases} \tag{12.10}$$

Example 12.5. Suppose that $b_1 = 0$, $b_2 = 45$, $\alpha_1 = 0.95$ and $\alpha_2 = 0.90$. A run of the hybrid intelligent algorithm (10000 cycles in fuzzy simulation, 1000 generations in GA) shows that the optimal schedule is

Machine 1: $6 \to 4$;
Machine 2: $1 \to 2 \to 3$;
Machine 3: $5 \to 8 \to 7$

which can satisfy the first goal, but the second objective is 0.12.

Fuzzy Dependent-Chance Goal Programming

Suppose that the management goals have the following priority structure.

At the first priority level, the credibility that the tardiness $f_1(\boldsymbol{x}, \boldsymbol{y}, \boldsymbol{\xi})$ does not exceed the given value b_1 should achieve a confidence level α_1. Thus we have a goal constraint

$$\alpha_1 - \text{Cr}\left\{f_1(\boldsymbol{x}, \boldsymbol{y}, \boldsymbol{\xi}) \le b_1\right\} = d_1^-$$

in which $d_1^- \vee 0$ will be minimized.

At the second priority level, the credibility that the makespan $f_2(\boldsymbol{x}, \boldsymbol{y}, \boldsymbol{\xi})$ does not exceed the given value b_2 should achieve a confidence level α_2. Thus we have a goal constraint

$$\alpha_2 - \text{Cr}\left\{f_2(\boldsymbol{x}, \boldsymbol{y}, \boldsymbol{\xi}) \le b_2\right\} = d_2^-$$

in which $d_2^- \vee 0$ will be minimized.

Then Peng and Liu [255] proposed the following fuzzy DCGP model for parallel machine scheduling problem,

$$\begin{cases} \text{lexmin}\{d_1^- \vee 0, d_2^- \vee 0\} \\ \text{subject to:} \\ \quad \alpha_1 - \text{Cr}\left\{f_1(\boldsymbol{x}, \boldsymbol{y}, \boldsymbol{\xi}) \le b_1\right\} = d_1^- \\ \quad \alpha_2 - \text{Cr}\left\{f_2(\boldsymbol{x}, \boldsymbol{y}, \boldsymbol{\xi}) \le b_2\right\} = d_2^- \\ \quad 1 \le x_i \le n, \quad i = 1, 2, \cdots, n \\ \quad x_i \ne x_j, \quad i \ne j, \ i, j = 1, 2, \cdots, n \\ \quad 0 \le y_1 \le y_2 \cdots \le y_{m-1} \le n \\ \quad x_i, y_j, \quad i = 1, 2, \cdots, n, \quad j = 1, 2, \cdots, m-1, \quad \text{integers.} \end{cases} \tag{12.11}$$

Example 12.6. Suppose that $b_1 = 0$, $b_2 = 45$, $\alpha_1 = 0.95$ and $\alpha_2 = 0.90$. A run of the hybrid intelligent algorithm (10000 cycles in fuzzy simulation, 1000 generations in GA) shows that the optimal schedule is

Machine 1: $1 \to 5 \to 3$;
Machine 2: $6 \to 4 \to 2$;
Machine 3: $8 \to 7$

which can satisfy the first goal, but the second objective is 0.05.

12.4 Hybrid Models

We assume that the processing times are hybrid variables, and introduce some hybrid models for machine scheduling problem.

Hybrid Expected Time Goal Programming

In order to balance the conflicting objectives, we may have the following target levels and priority structure:

At the first priority level, the expected tardiness $E[f_1(\boldsymbol{x},\boldsymbol{y},\boldsymbol{\xi})]$ should not exceed the target value b_1. Thus we have a goal constraint

$$E[f_1(\boldsymbol{x},\boldsymbol{y},\boldsymbol{\xi})] - b_1 = d_1^+$$

in which $d_1^+ \vee 0$ will be minimized.

At the second priority level, the expected makespan $E[f_2(\boldsymbol{x},\boldsymbol{y},\boldsymbol{\xi})]$ should not exceed the target value b_2. That is, we have a goal constraint

$$E[f_2(\boldsymbol{x},\boldsymbol{y},\boldsymbol{\xi})] - b_2 = d_2^+$$

in which $d_2^+ \vee 0$ will be minimized.

Then we have the following hybrid expected time goal programming model for the parallel machine scheduling problem,

$$\begin{cases} \text{lexmin}\left\{d_1^+ \vee 0, d_2^+ \vee 0\right\} \\ \text{subject to:} \\ \qquad E[f_1(\boldsymbol{x},\boldsymbol{y},\boldsymbol{\xi})] - b_1 = d_1^+ \\ \qquad E[f_2(\boldsymbol{x},\boldsymbol{y},\boldsymbol{\xi})] - b_2 = d_2^+ \\ \qquad 1 \le x_i \le n, \quad i = 1,2,\cdots,n \\ \qquad x_i \ne x_j, \quad i \ne j,\ i,j = 1,2,\cdots,n \\ \qquad 0 \le y_1 \le y_2 \cdots \le y_{m-1} \le n \\ \qquad x_i, y_j, \quad i = 1,2,\cdots,n, \quad j = 1,2,\cdots,m-1, \quad \text{integers.} \end{cases} \tag{12.12}$$

Hybrid Chance-Constrained Goal Programming

We assume the following priority structure. At the first priority level, the tardiness $f_1(\boldsymbol{x},\boldsymbol{y},\boldsymbol{\xi})$ should not exceed the target value b_1 with confidence level α_1. Thus we have a goal constraint

$$\text{Ch}\left\{f_1(\boldsymbol{x},\boldsymbol{y},\boldsymbol{\xi}) - b_1 \le d_1^+\right\} \ge \alpha_1$$

in which $d_1^+ \vee 0$ will be minimized.

At the second priority level, the makespan $f_2(\boldsymbol{x},\boldsymbol{y},\boldsymbol{\xi})$ should not exceed the target value b_2 with confidence level α_2. Thus we have a goal constraint

$$\text{Ch}\left\{f_2(\boldsymbol{x},\boldsymbol{y},\boldsymbol{\xi})-b_2\le d_2^+\right\}\ge\alpha_2$$

in which $d_2^+ \vee 0$ will be minimized.

Then we have the following hybrid CCGP model for the parallel machine scheduling problem,

$$\begin{cases}\text{lexmin}\{d_1^+ \vee 0, d_2^+ \vee 0\}\\ \text{subject to:}\\ \quad \text{Ch}\left\{f_1(\boldsymbol{x},\boldsymbol{y},\boldsymbol{\xi})-b_1\le d_1^+\right\}\ge\alpha_1\\ \quad \text{Ch}\left\{f_2(\boldsymbol{x},\boldsymbol{y},\boldsymbol{\xi})-b_2\le d_2^+\right\}\ge\alpha_2\\ \quad 1\le x_i\le n,\quad i=1,2,\cdots,n\\ \quad x_i\ne x_j,\quad i\ne j,\ i,j=1,2,\cdots,n\\ \quad 0\le y_1\le y_2\cdots\le y_{m-1}\le n\\ \quad x_i,\ y_j,\quad i=1,2,\cdots,n,\quad j=1,2,\cdots,m-1,\quad \text{integers.}\end{cases}\tag{12.13}$$

Hybrid Dependent-Chance Goal Programming

Suppose that the management goals have the following priority structure.

At the first priority level, the chance that the tardiness $f_1(\boldsymbol{x},\boldsymbol{y},\boldsymbol{\xi})$ does not exceed the given value b_1 should achieve a confidence level β_1. Thus we have a goal constraint

$$\beta_1-\text{Ch}\left\{f_1(\boldsymbol{x},\boldsymbol{y},\boldsymbol{\xi})\le b_1\right\}=d_1^-$$

in which $d_1^- \vee 0$ will be minimized.

At the second priority level, the chance that the makespan $f_2(\boldsymbol{x},\boldsymbol{y},\boldsymbol{\xi})$ does not exceed the given value b_2 should achieve a confidence level β_2. Thus we have a goal constraint

$$\beta_2-\text{Ch}\left\{f_2(\boldsymbol{x},\boldsymbol{y},\boldsymbol{\xi})\le b_2\right\}=d_2^-$$

in which $d_2^- \vee 0$ will be minimized.

Then we have the following hybrid DCGP model for parallel machine scheduling problem,

$$\begin{cases}\text{lexmin}\{d_1^- \vee 0, d_2^- \vee 0\}\\ \text{subject to:}\\ \quad \beta_1-\text{Ch}\left\{f_1(\boldsymbol{x},\boldsymbol{y},\boldsymbol{\xi})\le b_1\right\}=d_1^-\\ \quad \beta_2-\text{Ch}\left\{f_2(\boldsymbol{x},\boldsymbol{y},\boldsymbol{\xi})\le b_2\right\}=d_2^-\\ \quad 1\le x_i\le n,\quad i=1,2,\cdots,n\\ \quad x_i\ne x_j,\quad i\ne j,\ i,j=1,2,\cdots,n\\ \quad 0\le y_1\le y_2\cdots\le y_{m-1}\le n\\ \quad x_i,y_j,\quad i=1,2,\cdots,n,\quad j=1,2,\cdots,m-1,\quad \text{integers.}\end{cases}\tag{12.14}$$

12.5 Exercises

Problem 12.1. Design a hybrid intelligent algorithm to solve hybrid models for machine scheduling problem (for example, the processing times are random and the due dates are fuzzy).

Problem 12.2. Build uncertain models for machine scheduling problem (for example, the processing times are uncertain variables with identification function (λ, ρ)), and design a hybrid intelligent algorithm to solve them.

References

[1] Abe, S.: Neural Networks and Fuzzy Systems: Theory and Applications. Kluwer Academic Publishers, Boston (1997)

[2] Aggarwal, K.K., Chopra, Y.C., Bajwa, J.S.: Topological layout of links for optimizing the s-t reliability in a computer communication system. Microelectronics & Reliability 22(3), 341–345 (1982)

[3] Aggarwal, K.K., Chopra, Y.C., Bajwa, J.S.: Topological layout of links for optimizing the overall reliability in a computer communication system. Microelectronics & Reliability 22(3), 347–351 (1982)

[4] Angelov, P.: A generalized approach to fuzzy optimization. International Journal of Intelligent Systems 9, 261–268 (1994)

[5] Armentano, V.A., Ronconi, D.P.: Tabu search for total tardiness minimization in flowshop scheduling problems. Computers and Operations Research 26(3), 219–235 (1999)

[6] Ballestero, E.: Using stochastic goal programming: Some applications to management and a case of industrial production. INFOR 43(2), 63–77 (2005)

[7] Bandemer, H., Nather, W.: Fuzzy Data Analysis. Kluwer Academic Publishers, Dordrecht (1992)

[8] Bard, J.F.: An algorithm for solving the general bilevel programming problem. Mathematics of Operations Research 8, 260–272 (1983)

[9] Bard, J.F., Moore, J.T.: A branch and bound algorithm for the bilevel programming problem. SIAM J. Sci. Statist. Comput. 11, 281–292 (1990)

[10] Bard, J.F.: Optimality conditions for the bilevel programming problem. Naval Research Logistics Quarterly 31, 13–26 (1984)

[11] Bastian, C., Rinnooy Kan, A.H.G.: The stochastic vehicle routing problem revisited. European Journal of Operational Research 56, 407–412 (1992)

[12] Bellman, R.E.: Dynamic Programming. Princeton University Press, New Jersey (1957)

[13] Bellman, R.E., Zadeh, L.A.: Decision making in a fuzzy environment. Management Science 17, 141–164 (1970)

[14] Ben-Ayed, O., Blair, C.E.: Computational difficulties of bilevel linear programming. Operations Research 38, 556–560 (1990)

[15] Ben Abdelaziz, F., Enneifar, L., Martel, J.M.: A multiobjective fuzzy stochastic program for water resources optimization: The case of lake management. INFOR 42(3), 201–214 (2004)

[16] Bertsekas, D.P., Tsitsiklis, J.N.: Parallel and Distributed Computation: Numerical Methods. Prentice-Hall, Englewood Cliffs (1989)
[17] Bertsimas, D.J., Simchi-Levi, D.: A new generation of vehicle routing research: Robust algorithms, addressing uncertainty. Operations Research 44(2), 286–304 (1996)
[18] Bialas, W.F., Karwan, M.H.: Two-level linear programming. Management Science 30, 1004–1020 (1984)
[19] Biswal, M.P.: Fuzzy programming technique to solve multi-objective geometric programming problems. Fuzzy Sets and Systems 51, 61–71 (1992)
[20] Bit, A.K., Biswal, M.P., Alam, S.S.: Fuzzy programming approach to multicriteria decision making transportation problem. Fuzzy Sets and Systems 50, 135–141 (1992)
[21] Bitran, G.R.: Linear multiple objective problems with interval coefficients. Management Science 26, 694–706 (1980)
[22] Bouchon-Meunier, B., Kreinovich, V., Lokshin, A., Nguyen, H.T.: On the formulation of optimization under elastic constraints (with control in mind). Fuzzy Sets and Systems 81, 5–29 (1996)
[23] Bratley, P., Fox, B.L., Schrage, L.E.: A Guide to Simulation. Springer, New York (1987)
[24] Brucker, P.: Scheduling Algorithms, 2nd edn. Springer, New York (1998)
[25] Buckley, J.J.: Possibility and necessity in optimization. Fuzzy Sets and Systems 25, 1–13 (1988)
[26] Buckley, J.J.: Stochastic versus possibilistic programming. Fuzzy Sets and Systems 34, 173–177 (1990)
[27] Buckley, J.J.: Multiobjective possibilistic linear programming. Fuzzy Sets and Systems 35, 23–28 (1990)
[28] Buckley, J.J., Hayashi, Y.: Fuzzy genetic algorithm and applications. Fuzzy Sets and Systems 61, 129–136 (1994)
[29] Buckley, J.J., Feuring, T.: Evolutionary algorithm solution to fuzzy problems: Fuzzy linear programming. Fuzzy Sets and Systems 109(1), 35–53 (2000)
[30] Cadenas, J.M., Verdegay, J.L.: Using fuzzy numbers in linear programming. IEEE Transactions on Systems, Man and Cybernetics–Part B 27(6), 1016–1022 (1997)
[31] Campos, L., González, A.: A subjective approach for ranking fuzzy numbers. Fuzzy Sets and Systems 29, 145–153 (1989)
[32] Campos, L., Verdegay, J.L.: Linear programming problems and ranking of fuzzy numbers. Fuzzy Sets and Systems 32, 1–11 (1989)
[33] Campos, F.A., Villar, J., Jimenez, M.: Robust solutions using fuzzy chance constraints. Engineering Optimization 38(6), 627–645 (2006)
[34] Candler, W., Townsley, R.: A linear two-level programming problem. Computers and Operations Research 9, 59–76 (1982)
[35] Carlsson, C., Fuller, R., Majlender, P.: A possibilistic approach to selecting portfolios with highest utility score. Fuzzy Sets and Systems 131(1), 13–21 (2002)
[36] Changchit, C., Terrell, M.P.: CCGP model for multiobjective reservoir systems. Journal of Water Resources Planning and Management 115(5), 658–670 (1989)
[37] Charnes, A., Cooper, W.W.: Chance-constrained programming. Management Science 6(1), 73–79 (1959)

[38] Charnes, A., Cooper, W.W.: Management Models and Industrial Applications of Linear Programming. Wiley, New York (1961)

[39] Charnes, A., Cooper, W.W.: Chance constraints and normal deviates. Journal of the American Statistical Association 57, 134–148 (1962)

[40] Charnes, A., Cooper, W.W.: Deterministic equivalents for optimizing and satisficing under chance-constraints. Operations Research 11(1), 18–39 (1963)

[41] Charnes, A., Cooper, W.W.: Goal programming and multiple objective optimizations: Part I. European Journal of Operational Research 1(1), 39–54 (1977)

[42] Chanas, S.: Fuzzy programming in multiobjective linear programming—a parametric approach. Fuzzy Sets and Systems 29, 303–313 (1989)

[43] Chanas, S., Kuchta, D.: Multiobjective programming in optimization of interval objective functions—a generalized approach. European Journal of Operational Research 94, 594–598 (1996)

[44] Chen, A., Ji, Z.W.: Path finding under uncertainty. Journal of Advance Transportation 39(1), 19–37 (2005)

[45] Chen, S.J., Hwang, C.L.: Fuzzy Multiple Attribute Decision Making: Methods and Applications. Springer, Berlin (1992)

[46] Chen, Y., Fung, R.Y.K., Yang, J.: Fuzzy expected value modelling approach for determining target values of engineering characteristics in QFD. International Journal of Production Research 43(17), 3583–3604 (2005)

[47] Chen, Y., Fung, R.Y.K., Tang, J.F.: Rating technical attributes in fuzzy QFD by integrating fuzzy weighted average method and fuzzy expected value operator. European Journal of Operational Research 174(3), 1553–1566 (2006)

[48] Chopra, Y.C., Sohi, B.S., Tiwari, R.K., Aggarwal, K.K.: Network topology for maximizing the terminal reliability in a computer communication network. Microelectronics & Reliability 24, 911–913 (1984)

[49] Clayton, E., Weber, W., Taylor III, B.: A goal programming approach to the optimization of multiresponse simulation models. IIE Trans. 14, 282–287 (1982)

[50] Cooper, L.: Location-allocation problems. Operations Research 11, 331–344 (1963)

[51] Cybenko, G.: Approximations by superpositions of a sigmoidal function. Mathematics of Control, Signals and Systems 2, 183–192 (1989)

[52] Dantzig, G.B.: Linear Programming and Extensions. Princeton University Press, Princeton (1963)

[53] Das, B., Maity, K., Maiti, A.: A two warehouse supply-chain model under possibility/necessity/credibility measures. Mathematical and Computer Modelling 46(3-4), 398–409 (2007)

[54] Delgado, M., Verdegay, J.L., Vila, M.A.: A general model for fuzzy linear programming. Fuzzy Sets and Systems 29, 21–29 (1989)

[55] Dengiz, B., Altiparmak, F., Smith, A.E.: Efficient optimization of all-terminal reliable networks using an evolutionary approach. IEEE Transactions on Reliability 46(1), 18–26 (1997)

[56] Dror, M., Laporte, G., Trudreau, P.: Vehicle routing with stochastic demands: Properties and solution frameworks. Transportation Science 23, 166–176 (1989)

[57] Dubois, D., Prade, H.: Fuzzy Sets and Systems. Theory and Applications. Academic Press, New York (1980)

[58] Dubois, D., Prade, H.: Fuzzy cardinality and the modeling of imprecise quantification. Fuzzy Sets and Systems 16, 199–230 (1985)

[59] Dubois, D., Prade, H.: The mean value of a fuzzy number. Fuzzy Sets and Systems 24, 279–300 (1987)

[60] Dubois, D., Prade, H.: Possibility Theory: An Approach to Computerized Processing of Uncertainty. Plenum, New York (1988)

[61] Dunyak, J., Saad, I.W., Wunsch, D.: A theory of independent fuzzy probability for system reliability. IEEE Transactions on Fuzzy Systems 7(3), 286–294 (1999)

[62] Ermoliev, Y., Wets, R.J.B. (eds.): Numerical Techniques for Stochastic Optimization. Springer, Berlin (1988)

[63] Esogbue, A.O., Liu, B.: Reservoir operations optimization via fuzzy criterion decision processes. Fuzzy Optimization and Decision Making 5(3), 289–305 (2006)

[64] Feng, X., Liu, Y.K.: Measurability criteria for fuzzy random vectors. Fuzzy Optimization and Decision Making 5(3), 245–253 (2006)

[65] Feng, Y., Yang, L.X.: A two-objective fuzzy k-cardinality assignment problem. Journal of Computational and Applied Mathematics 197(1), 233–244 (2006)

[66] Fine, T.L.: Feedforward Neural Network Methodology. Springer, New York (1999)

[67] Fishman, G.S.: Monte Carlo: Concepts, Algorithms, and Applications. Springer, New York (1996)

[68] Fogel, D.B.: An introduction to simulated evolutionary optimization. IEEE Transactions on Neural Networks 5, 3–14 (1994)

[69] Fogel, D.B.: Evolution Computation: Toward a New Philosophy of Machine Intelligence. IEEE Press, Piscataway (1995)

[70] Fortemps, P.: Jobshop scheduling with imprecise durations: A fuzzy approach. IEEE Transactions on Fuzzy Systems 5(4), 557–569 (1997)

[71] Freeman, R.J.: A generalized PERT. Operations Research 8(2), 281 (1960)

[72] Freeman, R.J.: A generalized network approach to project activity sequencing. IRE Transactions on Engineering Management 7(3), 103–107 (1960)

[73] Fuller, R.: On stability in fuzzy linear programming problems. Fuzzy Sets and Systems 30, 339–344 (1989)

[74] Fung, K.: A philosophy for allocating component reliabilities in a network. IEEE Transactions on Reliability 34(2), 151–153 (1985)

[75] Fung, R.Y.K., Chen, Y.Z., Chen, L.: A fuzzy expected value-based goal programing model for product planning using quality function deployment. Engineering Optimization 37(6), 633–647 (2005)

[76] Gao, J., Liu, B.: New primitive chance measures of fuzzy random event. International Journal of Fuzzy Systems 3(4), 527–531 (2001)

[77] Gao, J., Liu, B., Gen, M.: A hybrid intelligent algorithm for stochastic multilevel programming. IEE J. Transactions on Electronics, Information and Systems 124-C(10), 1991–1998 (2004)

[78] Gao, J., Liu, B.: Fuzzy multilevel programming with a hybrid intelligent algorithm. Computers & Mathematics with Applications 49, 1539–1548 (2005)

[79] Gao, J., Lu, M.: Fuzzy quadratic minimum spanning tree problem. Applied Mathematics and Computation 164(3), 773–788 (2005)

[80] Gao, J., Feng, X.Q.: A hybrid intelligent algorithm for fuzzy dynamic inventory problem. Journal of Information and Computing Science 1(4), 235–244 (2006)

[81] Gao, J.: Credibilistic game with fuzzy information. Journal of Uncertain Systems 1(1), 74–80 (2007)
[82] Gen, M., Liu, B.: Evolution program for production plan problem. Engineering Design and Automation 1(3), 199–204 (1995)
[83] Gen, M., Liu, B., Ida, K.: Evolution program for deterministic and stochastic optimizations. European Journal of Operational Research 94(3), 618–625 (1996)
[84] Gen, M., Liu, B.: Evolution program for optimal capacity expansion. Journal of Operations Research Society of Japan 40(1), 1–9 (1997)
[85] Gen, M., Cheng, R.: Genetic Algorithms & Engineering Design. Wiley, New York (1997)
[86] Gen, M., Liu, B.: A genetic algorithm for nonlinear goal programming. Evolutionary Optimization 1(1), 65–76 (1999)
[87] Gen, M., Cheng, R.: Genetic Algorithms & Engineering Optimization. Wiley, New York (2000)
[88] Gendreau, M., Laporte, G., Séguin, R.: Stochastic vehicle routing. European Journal of Operational Research 88, 3–12 (1999)
[89] Goldberg, D.E.: Genetic Algorithms in Search, Optimization and Machine Learning. Addison-Wesley, Reading (1989)
[90] González, A.: A study of the ranking function approach through mean values. Fuzzy Sets and Systems 35, 29–41 (1990)
[91] Han, S., Ishii, H., Fujii, S.: One machine scheduling problem with fuzzy duedates. European Journal of Operational Research 79(1), 1–12 (1994)
[92] Hayashi, H., Buckley, J.J., Czogala, E.: Fuzzy neural network with fuzzy signals and weights. International Journal of Intelligent Systems 8, 527–537 (1993)
[93] He, Y., Xu, J.: A class of random fuzzy programming model and its application to vehicle routing problem. World Journal of Modelling and Simulation 1(1), 3–11 (2005)
[94] Heilpern, S.: The expected value of a fuzzy number. Fuzzy Sets and Systems 47, 81–86 (1992)
[95] Hisdal, E.: Logical Structures for Representation of Knowledge and Uncertainty. Physica-Verlag, Heidelberg (1998)
[96] Holland, J.H.: Adaptation in Natural and Artificial Systems. University of Michigan Press, Ann Arbor (1975)
[97] Hornik, K., Stinchcombe, M., White, H.: Multilayer feedforward networks are universal approximators. Neural Networks 2, 359–366 (1989)
[98] Hu, C.F., Fang, S.C.: Solving fuzzy inequalities with piecewise linear membership functions. IEEE Transactions on Fuzzy Systems 7, 230–235 (1999)
[99] Inuiguchi, M., Ichihashi, H., Kume, Y.: Relationships between modality constrained programming problems and various fuzzy mathematical programming problems. Fuzzy Sets and Systems 49, 243–259 (1992)
[100] Inuiguchi, M., Ichihashi, H., Kume, Y.: Modality constrained programming problems: An unified approach to fuzzy mathematical programming problems in the setting of possibility theory. Information Sciences 67, 93–126 (1993)
[101] Inuiguchi, M., Ramík, J.: Possibilistic linear programming: A brief review of fuzzy mathematical programming and a comparison with stochastic programming in portfolio selection problem. Fuzzy Sets and Systems 111(1), 3–28 (2000)

[102] Ishibuchi, H., Tanaka, H.: Multiobjective programming in optimization of the interval objective function. European Journal of Operational Research 48, 219–225 (1990)
[103] Ishibuchi, H., Tanaka, H.: An architecture of neural networks with interval weights and its application to fuzzy regression analysis. Fuzzy Sets and Systems 57, 27–39 (1993)
[104] Ishibuchi, H., Yamamoto, N., Murata, T., Tanaka, H.: Genetic algorithms and neighborhood search algorithms for fuzzy flowshop scheduling problems. Fuzzy Sets and Systems 67(1), 81–100 (1994)
[105] Iwamura, K., Liu, B.: A genetic algorithm for chance constrained programming. Journal of Information & Optimization Sciences 17(2), 409–422 (1996)
[106] Iwamura, K., Liu, B.: Chance constrained integer programming models for capital budgeting in fuzzy environments. Journal of the Operational Research Society 49(8), 854–860 (1998)
[107] Iwamura, K., Liu, B.: Stochastic operation models for open inventory networks. Journal of Information & Optimization Sciences 20(3), 347–363 (1999)
[108] Iwamura, K., Liu, B.: Dependent-chance integer programming applied to capital budgeting. Journal of the Operations Research Society of Japan 42(2), 117–127 (1999)
[109] Jan, R.H., Hwang, F.J., Chen, S.T.: Topological optimization of a communication network subject to a reliability constraint. IEEE Transactions on Reliability 42, 63–70 (1993)
[110] Ji, X.Y.: Models and algorithm for stochastic shortest path problem. Applied Mathematics and Computation 170(1), 503–514 (2005)
[111] Ji, X.Y., Shao, Z.: Model and algorithm for bilevel newsboy problem with fuzzy demands and discounts. Applied Mathematics and Computation 172(1), 163–174 (2006)
[112] Ji, X.Y., Iwamura, K., Shao, Z.: New models for shortest path problem with fuzzy arc lengths. Applied Mathematical Modelling 31(2), 259–269 (2007)
[113] Jiménez, F., Verdegay, J.L.: Uncertain solid transportation problems. Fuzzy Sets and Systems 100(1-3), 45–57 (1998)
[114] Kacprzyk, J., Esogbue, A.O.: Fuzzy dynamic programming: Main developments and applications. Fuzzy Sets and Systems 81, 31–45 (1996)
[115] Kacprzyk, J.: Multisatge control of a stochastic system in a fuzzy environment using a genetic algorithm. International Journal of Intelligent Systems 13, 1011–1023 (1998)
[116] Kacprzyk, J., Romero, R.A., Gomide, F.A.C.: Involving objective and subjective aspects in multistage decision making and control under fuzziness: Dynamic programming and neural networks. International Journal of Intelligent Systems 14, 79–104 (1999)
[117] Kacprzyk, J., Pasi, G., Vojtas, P.: Fuzzy querying: Issues and perspectives. Kybernetika 36, 605–616 (2000)
[118] Kacprzyk, J., Zadrozny, S.: Computing with words in intelligent database querying: Standalone and Internet-based applications. Information Sciences 134, 71–109 (2001)
[119] Kall, P., Wallace, S.W.: Stochastic Programming. Wiley, Chichester (1994)
[120] Kaufmann, A.: Introduction to the Theory of Fuzzy Subsets, vol. I. Academic Press, New York (1975)
[121] Kaufman, A., Gupta, M.M.: Introduction to Fuzzy Arithmetic: Theory and Applications. Van Nostrand Reinhold, New York (1985)

[122] Kaufman, A., Gupta, M.M.: Fuzzy Mathematical Models in Engineering and Management Science, 2nd edn. North-Holland, Amsterdam (1991)
[123] Ke, H., Liu, B.: Project scheduling problem with stochastic activity duration times. Applied Mathematics and Computation 168(1), 342–353 (2005)
[124] Ke, H., Liu, B.: Project scheduling problem with mixed uncertainty of randomness and fuzziness. European Journal of Operational Research 183(1), 135–147 (2007)
[125] Ke, H., Liu, B.: Fuzzy project scheduling problem and its hybrid intelligent algorithm, Technical Report (2008)
[126] Keller, J.M., Tahani, H.: Implementation of conjunctive and disjunctive fuzzy logic rules with neural networks. International Journal of Approximate Reasoning 6, 221–240 (1992)
[127] Keown, A.J.: A chance-constrained goal programming model for bank liquidity management. Decision Sciences 9, 93–106 (1978)
[128] Keown, A.J., Taylor, B.W.: A chance-constrained integer goal programming model for capital budgeting in the production area. Journal of the Operational Research Society 31(7), 579–589 (1980)
[129] Kitano, H.: Neurogenetic learning: An integrated method of designing and training neural networks using genetic algorithms. Physica D 75, 225–228 (1994)
[130] Klein, G., Moskowitz, H., Ravindran, A.: Interactive multiobjective optimization under uncertainty. Management Science 36(1), 58–75 (1990)
[131] Klement, E.P., Puri, M.L., Ralescu, D.A.: Limit theorems for fuzzy random variables. Proceedings of the Royal Society of London 407, 171–182 (1986)
[132] Klir, G.J., Folger, T.A.: Fuzzy Sets, Uncertainty, and Information. Prentice-Hall, Englewood Cliffs (1980)
[133] Klir, G.J., Yuan, B.: Fuzzy Sets and Fuzzy Logic: Theory and Applications. Prentice-Hall, New Jersey (1995)
[134] Kolonko, M.: Some new results on simulated annealing applied to the job shop scheduling problem. European Journal of Operational Research 113(1), 123–136 (1999)
[135] Konno, T., Ishii, H.: An open shop scheduling problem with fuzzy allowable time and fuzzy resource constraint. Fuzzy Sets and Systems 109(1), 141–147 (2000)
[136] Kou, W., Prasad, V.R.: An annotated overview of system-reliability optimization. IEEE Transactions on Reliability 49, 176–197 (2000)
[137] Koza, J.R.: Genetic Programming. MIT Press, Cambridge (1992)
[138] Koza, J.R.: Genetic Programming, II. MIT Press, Cambridge (1994)
[139] Kruse, R., Meyer, K.D.: Statistics with Vague Data. D. Reidel Publishing Company, Dordrecht (1987)
[140] Kumar, A., Pathak, R.M., Gupta, Y.P.: Genetic-algorithm-based reliability optimization for computer network expansion. IEEE Transactions on Reliability 44, 63–72 (1995)
[141] Kumar, A., Pathak, R.M., Gupta, Y.P., Parsaei, H.R.: A Genetic algorithm for distributed system topology design. Computers and Industrial Engineering 28, 659–670 (1995)
[142] Kwakernaak, H.: Fuzzy random variables–I: Definitions and theorems. Information Sciences 15, 1–29 (1978)
[143] Kwakernaak, H.: Fuzzy random variables–II: Algorithms and examples for the discrete case. Information Sciences 17, 253–278 (1979)

[144] Lai, Y.J., Hwang, C.L.: A new approach to some possibilistic linear programming problems. Fuzzy Sets and Systems 49, 121–133 (1992)
[145] Lai, Y.J., Hwang, C.L.: Fuzzy Multiple Objective Decision Making: Methods and Applications. Springer, New York (1994)
[146] Laporte, G., Louveaux, F.V., Mercure, H.: The vehicle routing problem with stochastic travel times. Transportation Science 26, 161–170 (1992)
[147] Law, A.M., Kelton, W.D.: Simulation Modelling & Analysis, 2nd edn. McGraw-Hill, New York (1991)
[148] Lee, E.S., Li, R.J.: Fuzzy multiple objective programming and compromise programming with Pareto optimum. Fuzzy Sets and Systems 53, 275–288 (1993)
[149] Lee, E.S.: Fuzzy multiple level programming. Applied Mathematics and Computation 120, 79–90 (2001)
[150] Lee, K.H.: First Course on Fuzzy Theory and Applications. Springer, Berlin (2005)
[151] Lee, S.M.: Goal Programming for Decision Analysis. Auerbach, Philadelphia (1972)
[152] Lee, S.M., Olson, D.L.: A gradient algorithm for chance constrained nonlinear goal programming. European Journal of Operational Research 22, 359–369 (1985)
[153] Lertworasirkul, S., Fang, S.C., Joines, J.A., Nuttle, H.L.W.: Fuzzy data envelopment analysis (DEA): a possibility approach. Fuzzy Sets and Systems 139(2), 379–394 (2003)
[154] Li, H.L., Yu, C.S.: A fuzzy multiobjective program with quasiconcave membership functions and fuzzy coefficients. Fuzzy Sets and Systems 109(1), 59–81 (2000)
[155] Li, J., Xu, J., Gen, M.: A class of multiobjective linear programming model with fuzzy random coefficients. Mathematical and Computer Modelling 44(11-12), 1097–1113 (2006)
[156] Li, P., Liu, B.: Entropy of credibility distributions for fuzzy variables. IEEE Transactions on Fuzzy Systems 16(1), 123–129 (2008)
[157] Li, S.Y., Hu, W., Yang, Y.P.: Receding horizon fuzzy optimization under local information environment with a case study. International Journal of Information Technology & Decision Making 3(1), 109–127 (2004)
[158] Li, X., Liu, B.: The independence of fuzzy variables with applications. International Journal of Natural Sciences & Technology 1(1), 95–100 (2006)
[159] Li, X., Liu, B.: New independence definition of fuzzy random variable and random fuzzy variable. World Journal of Modelling and Simulation 2(5), 338–342 (2006)
[160] Li, X., Liu, B.: A sufficient and necessary condition for credibility measures. International Journal of Uncertainty, Fuzziness & Knowledge-Based Systems 14(5), 527–535 (2006)
[161] Li, X., Liu, B.: Chance measure for hybrid events with fuzziness and randomness. Soft Computing 13(2), 105–115 (2009)
[162] Li, X., Liu, B.: Conditional chance measure for hybrid events, Technical Report (2008)
[163] Litoiu, M., Tadei, R.: Real-time task scheduling with fuzzy deadlines and processing times. Fuzzy Sets and Systems 117(1), 35–45 (2001)
[164] Liu, B.: Dependent-chance goal programming and its genetic algorithm based approach. Mathematical and Computer Modelling 24(7), 43–52 (1996)

[165] Liu, B., Esogbue, A.O.: Fuzzy criterion set and fuzzy criterion dynamic programming. Journal of Mathematical Analysis and Applications 199(1), 293–311 (1996)
[166] Liu, B.: Dependent-chance programming: A class of stochastic optimization. Computers & Mathematics with Applications 34(12), 89–104 (1997)
[167] Liu, B., Iwamura, K.: Modelling stochastic decision systems using dependent-chance programming. European Journal of Operational Research 101(1), 193–203 (1997)
[168] Liu, B., Iwamura, K.: Chance constrained programming with fuzzy parameters. Fuzzy Sets and Systems 94(2), 227–237 (1998)
[169] Liu, B., Iwamura, K.: A note on chance constrained programming with fuzzy coefficients. Fuzzy Sets and Systems 100(1-3), 229–233 (1998)
[170] Liu, B.: Stackelberg-Nash equilibrium for multilevel programming with multiple followers using genetic algorithms. Computers & Mathematics with Applications 36(7), 79–89 (1998)
[171] Liu, B.: Minimax chance constrained programming models for fuzzy decision systems. Information Sciences 112(1-4), 25–38 (1998)
[172] Liu, B.: Dependent-chance programming with fuzzy decisions. IEEE Transactions on Fuzzy Systems 7(3), 354–360 (1999)
[173] Liu, B., Esogbue, A.O.: Decision Criteria and Optimal Inventory Processes. Kluwer Academic Publishers, Boston (1999)
[174] Liu, B.: Uncertain Programming. Wiley, New York (1999)
[175] Liu, B.: Dependent-chance programming in fuzzy environments. Fuzzy Sets and Systems 109(1), 97–106 (2000)
[176] Liu, B., Iwamura, K.: Topological optimization models for communication network with multiple reliability goals. Computers & Mathematics with Applications 39, 59–69 (2000)
[177] Liu, B.: Uncertain programming: A unifying optimization theory in various uncertain environments. Applied Mathematics and Computation 120(1-3), 227–234 (2001)
[178] Liu, B., Iwamura, K.: Fuzzy programming with fuzzy decisions and fuzzy simulation-based genetic algorithm. Fuzzy Sets and Systems 122(2), 253–262 (2001)
[179] Liu, B.: Fuzzy random chance-constrained programming. IEEE Transactions on Fuzzy Systems 9(5), 713–720 (2001)
[180] Liu, B.: Fuzzy random dependent-chance programming. IEEE Transactions on Fuzzy Systems 9(5), 721–726 (2001)
[181] Liu, B.: Theory and Practice of Uncertain Programming. Physica-Verlag, Heidelberg (2002)
[182] Liu, B.: Toward fuzzy optimization without mathematical ambiguity. Fuzzy Optimization and Decision Making 1(1), 43–63 (2002)
[183] Liu, B., Lai, K.K.: Stochastic programming models for vehicle routing problems. Asian Information-Science-Life 1(1), 13–28 (2002)
[184] Liu, B., Liu, Y.K.: Expected value of fuzzy variable and fuzzy expected value models. IEEE Transactions on Fuzzy Systems 10(4), 445–450 (2002)
[185] Liu, B.: Random fuzzy dependent-chance programming and its hybrid intelligent algorithm. Information Sciences 141(3-4), 259–271 (2002)
[186] Liu, B.: Uncertainty Theory. Springer, Berlin (2004)
[187] Liu, B.: A survey of credibility theory. Fuzzy Optimization and Decision Making 5(4), 387–408 (2006)

[188] Liu, B.: A survey of entropy of fuzzy variables. Journal of Uncertain Systems 1(1), 4–13 (2007)
[189] Liu, B.: Uncertainty Theory, 2nd edn. Springer, Berlin (2007)
[190] Liu, B.: Fuzzy process, hybrid process and uncertain process. Journal of Uncertain Systems 2(1), 3–16 (2008)
[191] Liu, B.: Some research problems in uncertainty theory. Journal of Uncertain Systems 3(1), 3–10 (2009)
[192] Liu, B.: Uncertainty Theory, 3rd edn., `http://orsc.edu.cn/liu/ut.pdf`
[193] Liu, L.Z., Li, Y.Z.: The fuzzy quadratic assignment problem with penalty: New models and genetic algorithm. Applied Mathematics and Computation 174(2), 1229–1244 (2006)
[194] Liu, X.W.: Measuring the satisfaction of constraints in fuzzy linear programming. Fuzzy Sets and Systems 122(2), 263–275 (2001)
[195] Liu, Y.H.: How to generate uncertain measures. In: Proceedings of Tenth National Youth Conference on Information and Management Sciences, Luoyang, August 3-7, 2008, pp. 23–26 (2008)
[196] Liu, Y.K., Liu, B.: Random fuzzy programming with chance measures defined by fuzzy integrals. Mathematical and Computer Modelling 36(4-5), 509–524 (2002)
[197] Liu, Y.K., Liu, B.: Fuzzy random programming problems with multiple criteria. Asian Information-Science-Life 1(3), 249–256 (2002)
[198] Liu, Y.K., Liu, B.: A class of fuzzy random optimization: Expected value models. Information Sciences 155(1-2), 89–102 (2003)
[199] Liu, Y.K., Liu, B.: Fuzzy random variables: A scalar expected value operator. Fuzzy Optimization and Decision Making 2(2), 143–160 (2003)
[200] Liu, Y.K., Liu, B.: Expected value operator of random fuzzy variable and random fuzzy expected value models. International Journal of Uncertainty, Fuzziness & Knowledge-Based Systems 11(2), 195–215 (2003)
[201] Liu, Y.K., Liu, B.: On minimum-risk problems in fuzzy random decision systems. Computers & Operations Research 32(2), 257–283 (2005)
[202] Liu, Y.K., Liu, B.: Fuzzy random programming with equilibrium chance constraints. Information Sciences 170, 363–395 (2005)
[203] Liu, Y.K.: Fuzzy programming with recourse. International Journal of Uncertainty. Fuzziness & Knowledge-Based Systems 13(4), 381–413 (2005)
[204] Liu, Y.K.: Convergent results about the use of fuzzy simulation in fuzzy optimization problems. IEEE Transactions on Fuzzy Systems 14(2), 295–304 (2006)
[205] Liu, Y.K., Wang, S.M.: On the properties of credibility critical value functions. Journal of Information and Computing Science 1(4), 195–206 (2006)
[206] Liu, Y.K., Gao, J.: The independence of fuzzy variables with applications to fuzzy random optimization. International Journal of Uncertainty, Fuzziness & Knowledge-Based Systems 15(Suppl. 2), 1–20 (2007)
[207] Liu, Y.K.: The approximation method for two-stage fuzzy random programming with recourse. IEEE Transactions on Fuzzy Systems 15(6), 1197–1208 (2007)
[208] Loetamonphong, J., Fang, S.C.: An efficient solution procedure for fuzzy relation equations with max-product composition. IEEE Transactions on Fuzzy Systems 7, 441–445 (1999)
[209] Logendran, R., Terrell, M.P.: Uncapacitated plant location-allocation problems with price sensitive stochastic demands. Computers and Operations Research 15(2), 189–198 (1988)

[210] Lu, M.: On crisp equivalents and solutions of fuzzy programming with different chance measures. Information: An International Journal 6(2), 125–133 (2003)
[211] Lucas, C., Araabi, B.N.: Generalization of the Dempster-Shafer Theory: A fuzzy-valued measure. IEEE Transactions on Fuzzy Systems 7(3), 255–270 (1999)
[212] Luhandjula, M.K.: Linear programming under randomness and fuzziness. Fuzzy Sets and Systems 10, 45–55 (1983)
[213] Luhandjula, M.K.: On possibilistic linear programming. Fuzzy Sets and Systems 18, 15–30 (1986)
[214] Luhandjula, M.K.: Fuzzy optimization: An appraisal. Fuzzy Sets and Systems 30, 257–282 (1989)
[215] Luhandjula, M.K.: Fuzziness and randomness in an optimization framework. Fuzzy Sets and Systems 77, 291–297 (1996)
[216] Luhandjula, M.K., Gupta, M.M.: On fuzzy stochastic optimization. Fuzzy Sets and Systems 81, 47–55 (1996)
[217] Luhandjula, M.K.: Optimisation under hybrid uncertainty. Fuzzy Sets and Systems 146(2), 187–203 (2004)
[218] Luhandjula, M.K.: Fuzzy stochastic linear programming: Survey and future research directions. European Journal of Operational Research 174(3), 1353–1367 (2006)
[219] Mabuchi, S.: An interpretation of membership functions and the properties of general probabilistic operators as fuzzy set operators. Fuzzy Sets and Systems 92(1), 31–50 (1997)
[220] Maiti, M.K., Maiti, M.A.: Fuzzy inventory model with two warehouses under possibility constraints. Fuzzy Sets and Systems 157(1), 52–73 (2006)
[221] Maleki, H.R., Tata, M., Mashinchi, M.: Linear programming with fuzzy variables. Fuzzy Sets and Systems 109(1), 21–33 (2000)
[222] Maniezzo, V.: Genetic evolution of the topology and weight distribution of neural networks. IEEE Transactions on Neural Networks 5(1), 39–53 (1994)
[223] Mareš, M.: Computation Over Fuzzy Quantities. CRC Press, Boca Raton (1994)
[224] Mareschal, B.: Stochastic multicriteria decision making and uncertainty. European Journal of Operational Research 26(1), 58–64 (1986)
[225] Martel, A., Price, W.: Stochastic programming applied to human resource planning. Journal of the Operational Research Society 32, 187–196 (1981)
[226] Masud, A., Hwang, C.: Interactive sequential goal programming. Journal of the Operational Research Society 32, 391–400 (1981)
[227] Marianov, V., Rios, M., Barros, F.J.: Allocating servers to facilities, when demand is elastic to travel and waiting times. RAIRO-Operations Research 39(3), 143–162 (2005)
[228] Medsker, L.R.: Hybrid Intelligent Systems. Kluwer Academic Publishers, Boston (1995)
[229] Mendel, J.M.: Uncertainty Rule-Based Fuzzy Logic Systems: Introduction and New Directions. Prentice-Hall, New Jersey (2001)
[230] Michalewicz, Z.: Genetic Algorithms + Data Structures = Evolution Programs, 3rd edn. Springer, Berlin (1996)
[231] Minsky, M., Papert, S.: Perceptrons. MIT Press, Cambridge (1969)
[232] Mitchell, M.: An Introduction to Genetic Algorithms. MIT Press, Cambridge (1996)

[233] Mohamed, R.H.: A chance-constrained fuzzy goal program. Fuzzy Sets and Systems 47, 183–186 (1992)
[234] Mohammed, W.: Chance constrained fuzzy goal programming with right-hand side uniform random variable coefficients. Fuzzy Sets and Systems 109(1), 107–110 (2000)
[235] Mohan, C., Nguyen, H.T.: An interactive satisfying method for solving multiobjective mixed fuzzy-stochastic programming problems. Fuzzy Sets and Systems 117, 95–111 (2001)
[236] Morgan, B.: Elements of Simulation. Chapman & Hall, London (1984)
[237] Morgan, D.R., Eheart, J.W., Valocchi, A.J.: Aquifer remediation design under uncertainty using a new chance constrained programming technique. Water Resources Research 29(3), 551–561 (1993)
[238] Mukherjee, S.P.: Mixed strategies in chance-constrained programming. Journal of the Operational Research Society 31, 1045–1047 (1980)
[239] Murat, C., Paschos, V.T.H.: The probabilistic longest path problem. Networks 33, 207–219 (1999)
[240] Murray, A.T., Church, R.L.: Applying simulated annealing to location-planning models. Journal of Heuristics 2(1), 31–53 (1996)
[241] Nahmias, S.: Fuzzy variables. Fuzzy Sets and Systems 1, 97–110 (1978)
[242] Negoita, C.V., Ralescu, D.: On fuzzy optimization. Kybernetes 6, 193–195 (1977)
[243] Negoita, C.V., Ralescu, D.: Simulation, Knowledge-based Computing, and Fuzzy Statistics. Van Nostrand Reinhold, New York (1987)
[244] Nguyen, V.H.: Fuzzy stochastic goal programming problems. European Journal of Operational Research 176(1), 77–86 (2007)
[245] Ohlemuller, M.: Tabu search for large location-allocation problems. Journal of the Operational Research Society 48(7), 745–750 (1997)
[246] Ostasiewicz, W.: A new approach to fuzzy programming. Fuzzy Sets and Systems 7, 139–152 (1982)
[247] Painton, L., Campbell, J.: Genetic algorithms in optimization of system reliability. IEEE Transactions on Reliability 44, 172–178 (1995)
[248] Pawlak, Z.: Rough sets. International Journal of Information and Computer Sciences 11(5), 341–356 (1982)
[249] Pawlak, Z.: Rough sets and fuzzy sets. Fuzzy Sets and Systems 17, 99–102 (1985)
[250] Pawlak, Z.: Rough Sets: Theoretical Aspects of Reasoning about Data. Kluwer Academic Publishers, Dordrecht (1991)
[251] Pawlak, Z., Slowinski, R.: Rough set approach to multi-attribute decision analysis. European Journal of Operational Research 72, 443–459 (1994)
[252] Pawlak, Z.: Rough set approach to knowledge-based decision support. European Journal of Operational Research 99, 48–57 (1997)
[253] Pedrycz, W.: Optimization schemes for decomposition of fuzzy relations. Fuzzy Sets and Systems 100, 301–325 (1998)
[254] Peng, J., Liu, B.: Stochastic goal programming models for parallel machine scheduling problems. Asian Information-Science-Life 1(3), 257–266 (2002)
[255] Peng, J., Liu, B.: Parallel machine scheduling models with fuzzy processing times. Information Sciences 166(1-4), 49–66 (2004)
[256] Peng, J., Liu, B.: A framework of birandom theory and optimization methods. Information: An International Journal 9(4), 629–640 (2006)

[257] Peng, J., Zhao, X.D.: Some theoretical aspects of the critical values of birandom variable. Journal of Information and Computing Science 1(4), 225–234 (2006)
[258] Peng, J., Liu, B.: Birandom variables and birandom programming. Computers & Industrial Engineering 53(3), 433–453 (2007)
[259] Puri, M.L., Ralescu, D.: Fuzzy random variables. Journal of Mathematical Analysis and Applications 114, 409–422 (1986)
[260] Raj, P.A., Kumer, D.N.: Ranking alternatives with fuzzy weights using maximizing set and minimizing set. Fuzzy Sets and Systems 105, 365–375 (1999)
[261] Ramer, A.: Conditional possibility measures. International Journal of Cybernetics and Systems 20, 233–247 (1989)
[262] Ramík, J.: Extension principle in fuzzy optimization. Fuzzy Sets and Systems 19, 29–35 (1986)
[263] Ramík, J., Rommelfanger, H.: Fuzzy mathematical programming based on some inequality relations. Fuzzy Sets and Systems 81, 77–88 (1996)
[264] Ravi, V., Murty, B.S.N., Reddy, P.J.: Nonequilibrium simulated annealing-algorithm applied to reliability optimization of complex systems. IEEE Transactions on Reliability 46, 233–239 (1997)
[265] Rommelfanger, H., Hanscheck, R., Wolfe, J.: Linear programming with fuzzy objectives. Fuzzy Sets and Systems 29, 31–48 (1989)
[266] Roubens, M., Teghem Jr., J.: Comparison of methodologies for fuzzy and stochastic multi-objective programming. Fuzzy Sets and Systems 42, 119–132 (1991)
[267] Roy, B.: Main sources of inaccurate determination, uncertainty and imprecision in decision models. Mathematical and Computer Modelling 12, 1245–1254 (1989)
[268] Rubinstein, R.Y.: Simulation and the Monte Carlo Method. Wiley, New York (1981)
[269] Saade, J.J.: Maximization of a function over a fuzzy domain. Fuzzy Sets and Systems 62, 55–70 (1994)
[270] Saber, H.M., Ravindran, A.: Nonlinear goal programming theory and practice: A survey. Computers and Operations Research 20, 275–291 (1993)
[271] Sakawa, M., Yano, H.: Feasibility and Pareto optimality for multiobjective nonlinear programming problems with fuzzy parameters. Fuzzy Sets and Systems 43, 1–15 (1991)
[272] Sakawa, M., Kato, K., Katagiri, H.: An interactive fuzzy satisficing method for multiobjective linear programming problems with random variable coefficients through a probability maximization model. Fuzzy Sets and Systems 146(2), 205–220 (2004)
[273] Sakawa, M., Nishizaki, I., Uemura, Y.: Interactive fuzzy programming for multi-level linear programming problems with fuzzy parameters. Fuzzy Sets and Systems 109(1), 3–19 (2000)
[274] Sakawa, M., Kubota, R.: Fuzzy programming for multiobjective job shop scheduling with fuzzy processing time and fuzzy duedate through genetic algorithms. European Journal of Operational Research 120(2), 393–407 (2000)
[275] Sakawa, M., Nishizaki, I., Uemura, Y.: Interactive fuzzy programming for two-level linear fractional programming problems with fuzzy parameters. Fuzzy Sets and Systems 115, 93–103 (2000)
[276] Sakawa, M., Nishizaki, I., Hitaka, M.: Interactive fuzzy programming for multi-level 0-1 programming problems with fuzzy parameters through genetic algorithms. Fuzzy Sets and Systems 117, 95–111 (2001)

[277] Savard, G., Gauvin, J.: The steepest descent direction for nonlinear bilevel programming problem. Operations Research Letters 15, 265–272 (1994)

[278] Schalkoff, R.J.: Artificial Neural Networks. McGraw-Hill, New York (1997)

[279] Schneider, M., Kandel, A.: Properties of the fuzzy expected value and the fuzzy expected interval in fuzzy environment. Fuzzy Sets and Systems 26, 373–385 (1988)

[280] Shao, Z., Ji, X.Y.: Fuzzy multi-product constraint newsboy problem. Applied Mathematics and Computation 180(1), 7–15 (2006)

[281] Sherali, H.D., Rizzo, T.P.: Unbalanced, capacitated p-median problems on a chain graph with a continuum of link demands. Networks 21(2), 133–163 (1991)

[282] Shih, H.S., Lai, Y.J., Lee, E.S.: Fuzzy approach for multilevel programming problems. Computers and Operations Research 23, 73–91 (1996)

[283] Shimizu, K., Ishizuka, Y., Bard, J.F.: Nondifferentiable and Two-level Mathematical Programming. Kluwer, Boston (1997)

[284] Shin, W.S., Ravindran, A.: Interactive multiple objective optimization: Survey I - continuous case. Computers and Operations Research 18(1), 97–114 (1991)

[285] Silva, E.F., Wood, R.K.: Solving a class of stochastic mixed-integer programs with branch and price. Mathematical Programming 108(2-3), 395–418 (2007)

[286] Slowinski, R., Teghem Jr., J.: Fuzzy versus stochastic approaches to multicriteria linear programming under uncertainty. Naval Research Logistics 35, 673–695 (1988)

[287] Slowinski, R., Stefanowski, J.: Rough classification in incomplete information systems. Mathematical and Computer Modelling 12, 1347–1357 (1989)

[288] Slowinski, R., Vanderpooten, D.: A generalized definition of rough approximations based on similarity. IEEE Transactions on Knowledge and Data Engineering 12(2), 331–336 (2000)

[289] Sommer, G., Pollatschek, M.A.: A fuzzy programming approach to an air pollution regulation problem. European Journal of Operational Research 10, 303–313 (1978)

[290] Soroush, H.M.: The most critical path in a PERT network. Journal of the Operational Research Society 45, 287–300 (1994)

[291] Steuer, R.E.: Algorithm for linear programming problems with interval objective function coefficients. Mathematics of Operational Research 6, 333–348 (1981)

[292] Steuer, R.E.: Multiple Criteria Optimization: Theory.Computation and Application. Wiley, New York (1986)

[293] Stewart Jr., W.R., Golden, B.L.: Stochastic vehicle routing: A comprehensive approach. European Journal of Operational Research 14, 371–385 (1983)

[294] Suykens, J.A.K., Vandewalle, J.P.L., De Moor, B.L.R.: Artificial Neural Networks for Modelling and Control of Non-Linear Systems. Kluwer Academic Publishers, Boston (1996)

[295] Szmidt, E., Kacprzyk, J.: Entropy for intuitionistic fuzzy sets. Fuzzy Sets and Systems 118, 467–477 (2001)

[296] Taha, H.A.: Operations Research: An Introduction. Macmillan, New York (1982)

[297] Tanaka, H., Asai, K.: Fuzzy linear programming problems with fuzzy numbers. Fuzzy Sets and Systems 13, 1–10 (1984)

[298] Tanaka, H., Asai, K.: Fuzzy solutions in fuzzy linear programming problems. IEEE Transactions on Systems, Man and Cybernetics 14, 325–328 (1984)
[299] Tanaka, H., Guo, P.: Possibilistic Data Analysis for Operations Research. Physica-Verlag, Heidelberg (1999)
[300] Tanaka, H., Guo, P., Zimmermann, H.J.: Possibility distribution of fuzzy decision variables obtained from possibilistic linear programming problems. Fuzzy Sets and Systems 113, 323–332 (2000)
[301] Teghem Jr., J., DuFrane, D., Kunsch, P.: STRANGE: An interactive method for multiobjective linear programming under uncertainty. European Journal Operational Research 26(1), 65–82 (1986)
[302] Turunen, E.: Mathematics Behind Fuzzy Logic. Physica-Verlag, Heidelberg (1999)
[303] Van Rooij, A.J.F., Jain, L.C., Johnson, R.P.: Neural Network Training Using Genetic Algorithms. World Scientific, Singapore (1996)
[304] Wagner, B.J., Gorelick, S.M.: Optimal ground water quality management under parameter uncertainty. Water Resources Research 23(7), 1162–1174 (1987)
[305] Wang, D.: An inexact approach for linear programming problems with fuzzy objective and resource. Fuzzy Sets and Systems 89, 61–68 (1998)
[306] Wang, G., Qiao, Z.: Linear programming with fuzzy random variable coefficients. Fuzzy Sets and Systems 57, 295–311 (1993)
[307] Wang, Y.P., Jiao, Y.C., Li, H.: An evolutionary algorithm for solving nonlinear bilevel programming based on a new constraint-handling scheme. IEEE Transactions on Systems, Man and Cybernetics Part C 35(2), 221–232 (2005)
[308] Waters, C.D.J.: Vehicle-scheduling problems with uncertainty and omitted customers. Journal of the Operational Research Society 40, 1099–1108 (1989)
[309] Weistroffer, H.: An interactive goal programming method for nonlinear multiple-criteria decision-making problems. Computers and Operations Research 10(4), 311–320 (1983)
[310] Wen, M., Iwamura, K.: Fuzzy facility location-allocation problem under the Hurwicz criterion. European Journal of Operational Research 184(2), 627–635 (2008)
[311] Wen, M., Iwamura, K.: Facility location-allocation problem in random fuzzy environment: Using (α, β)-cost minimization model under the Hurewicz criterion. Computers and Mathematics with Applications 55(4), 704–713 (2008)
[312] Whalen, T.: Decision making under uncertainty with various assumptions about available information. IEEE Transactions on Systems, Man and Cybernetics 14, 888–900 (1984)
[313] Xu, J., Li, J.: A class of stochastic optimization problems with one quadratic and several linear objective functions and extended portfolio selection model. Journal of Computational and Applied Mathematics 146(1), 99–113 (2002)
[314] Yager, R.R.: Mathematical programming with fuzzy constraints and a preference on the objective. Kybernetes 9, 285–291 (1979)
[315] Yager, R.R.: A procedure for ordering fuzzy subsets of the unit interval. Information Sciences 24, 143–161 (1981)
[316] Yager, R.R.: Modeling uncertainty using partial information. Information sciences 121, 271–294 (1999)
[317] Yager, R.R.: Decision making with fuzzy probability assessments. IEEE Transactions on Fuzzy Systems 7, 462–466 (1999)

[318] Yager, R.R.: On the evaluation of uncertain courses of action. Fuzzy Optimization and Decision Making 1, 13–41 (2002)
[319] Yang, L., Liu, B.: On inequalities and critical values of fuzzy random variable. International Journal of Uncertainty, Fuzziness & Knowledge-Based Systems 13(2), 163–175 (2005)
[320] Yang, L., Liu, B.: A sufficient and necessary condition for chance distribution of birandom variable. Information: An International Journal 9(1), 33–36 (2006)
[321] Yang, N., Wen, F.S.: A chance constrained programming approach to transmission system expansion planning. Electric Power Systems Research 75(2-3), 171–177 (2005)
[322] Yao, Y.Y.: Two views of the theory of rough sets in finite universes. International Journal of Approximate Reasoning 15, 291–317 (1996)
[323] Yazenin, A.V.: Fuzzy and stochastic programming. Fuzzy Sets and Systems 22, 171–180 (1987)
[324] Yazenin, A.V.: On the problem of possibilistic optimization. Fuzzy Sets and Systems 81, 133–140 (1996)
[325] Zadeh, L.A.: Fuzzy sets. Information and Control 8, 338–353 (1965)
[326] Zadeh, L.A.: Outline of a new approach to the analysis of complex systems and decision processes. IEEE Transactions on Systems, Man and Cybernetics 3, 28–44 (1973)
[327] Zadeh, L.A.: The concept of a linguistic variable and its application to approximate reasoning. Information Sciences 8, 199–251 (1975)
[328] Zadeh, L.A.: Fuzzy sets as a basis for a theory of possibility. Fuzzy Sets and Systems 1, 3–28 (1978)
[329] Zadeh, L.A.: A theory of approximate reasoning. In: Hayes, J., Michie, D., Thrall, R.M. (eds.) Mathematical Frontiers of the Social and Policy Science, pp. 69–129. Westview Press, Boulder (1979)
[330] Zhao, R., Liu, B.: Stochastic programming models for general redundancy optimization problems. IEEE Transactions on Reliability 52(2), 181–191 (2003)
[331] Zhao, R., Liu, B.: Renewal process with fuzzy interarrival times and rewards. International Journal of Uncertainty, Fuzziness & Knowledge-Based Systems 11(5), 573–586 (2003)
[332] Zhao, R., Liu, B.: Redundancy optimization problems with uncertainty of combining randomness and fuzziness. European Journal of Operational Research 157(3), 716–735 (2004)
[333] Zhao, R., Liu, B.: Standby redundancy optimization problems with fuzzy lifetimes. Computers & Industrial Engineering 49(2), 318–338 (2005)
[334] Zhao, R., Tang, W.S., Yun, H.L.: Random fuzzy renewal process. European Journal of Operational Research 169(1), 189–201 (2006)
[335] Zhao, R., Tang, W.S.: Some properties of fuzzy random renewal process. IEEE Transactions on Fuzzy Systems 14(2), 173–179 (2006)
[336] Zheng, Y., Liu, B.: Fuzzy vehicle routing model with credibility measure and its hybrid intelligent algorithm. Applied Mathematics and Computation 176(2), 673–683 (2006)
[337] Zhou, J., Liu, B.: New stochastic models for capacitated location-allocation problem. Computers & Industrial Engineering 45(1), 111–125 (2003)
[338] Zhou, J., Liu, B.: Analysis and algorithms of bifuzzy systems. International Journal of Uncertainty, Fuzziness & Knowledge-Based Systems 12(3), 357–376 (2004)

[339] Zhou, J., Liu, B.: Modeling capacitated location-allocation problem with fuzzy demands. Computers & Industrial Engineering 53(3), 454–468 (2007)
[340] Zhu, Y., Liu, B.: Continuity theorems and chance distribution of random fuzzy variable. Proceedings of the Royal Society of London Series A 460, 2505–2519 (2004)
[341] Zhu, Y., Liu, B.: Some inequalities of random fuzzy variables with application to moment convergence. Computers & Mathematics with Applications 50(5-6), 719–727 (2005)
[342] Zimmermann, H.J.: Description and optimization of fuzzy systems. International Journal of General Systems 2, 209–215 (1976)
[343] Zimmermann, H.J.: Fuzzy programming and linear programming with several objective functions. Fuzzy Sets and Systems 3, 45–55 (1978)
[344] Zimmermann, H.J.: Fuzzy mathematical programming. Computers and Operations Research 10, 291–298 (1983)
[345] Zimmermann, H.J.: Applications of fuzzy set theory to mathematical programming. Information Sciences 36, 29–58 (1985)
[346] Zimmermann, H.J.: Fuzzy Set Theory and its Applications. Kluwer Academic Publishers, Boston (1985)

List of Acronyms

CCDP	Chance-Constrained Dynamic Programming
CCGP	Chance-Constrained Goal Programming
CCMLP	Chance-Constrained Multilevel Programming
CCMOP	Chance-Constrained Multiobjective Programming
CCP	Chance-Constrained Programming
DCDP	Dependent-Chance Dynamic Programming
DCGP	Dependent-Chance Goal Programming
DCMLP	Dependent-Chance Multilevel Programming
DCMOP	Dependent-Chance Multiobjective Programming
DCP	Dependent-Chance Programming
DP	Dynamic Programming
EVDP	Expected Value Dynamic Programming
EVGP	Expected Value Goal Programming
EVM	Expected Value Model
EVMLP	Expected Value Multilevel Programming
EVMOP	Expected Value Multiobjective Programming
GA	Genetic Algorithm
GP	Goal Programming
MLP	Multilevel Programming
MOP	Multiobjective Programming
NN	Neural Network
SOP	Single-Objective Programming

List of Frequently Used Symbols

x, y, z	decision variables
$\boldsymbol{x}$, $\boldsymbol{y}$, $\boldsymbol{z}$	decision vectors
$\tilde{a}$, $\tilde{b}$, $\tilde{c}$	fuzzy variables
$\tilde{\boldsymbol{a}}$, $\tilde{\boldsymbol{b}}$, $\tilde{\boldsymbol{c}}$	fuzzy vectors
ξ, η, τ	random, fuzzy, or hybrid variables
$\boldsymbol{\xi}$, $\boldsymbol{\eta}$, $\boldsymbol{\tau}$	random, fuzzy, or hybrid vectors
μ, ν	membership functions
ϕ, ψ	probability density functions
Φ, Ψ	probability distributions
f, f_i	objective functions
g, g_j	constraint functions
$\emptyset$	empty set
Pr	probability measure
Cr	credibility measure
Ch	chance measure
$\mathcal{M}$	uncertain measure
E	expected value
$(\Omega, \mathcal{A}, \mathrm{Pr})$	probability space
$(\Theta, \mathcal{P}, \mathrm{Cr})$	credibility space
$(\Theta, \mathcal{P}, \mathrm{Cr}) \times (\Omega, \mathcal{A}, \mathrm{Pr})$	chance space
$(\Gamma, \mathcal{L}, \mathcal{M})$	uncertainty space
α, β	confidence levels
d^+, d^-	positive and negative deviations
$\Re$	set of real numbers
$\Re^n$	set of n-dimensional real vectors
$\vee$	maximum operator
$\wedge$	minimum operator
lexmin	lexicographical minimization

Index